ADVENTURES IN
CELESTIAL MECHANICS

ADVENTURES IN CELESTIAL MECHANICS

Second Edition

VICTOR G. SZEBEHELY
HANS MARK

Department of Aerospace Engineering
and Engineering Mechanics
The University of Texas at Austin

A Wiley-Interscience Publication
JOHN WILEY & SONS, INC.
New York / Chichester / Weinheim / Brisbane / Singapore / Toronto

Library of Congress Cataloging in Publication Data:

Szebehely, Victor G., 1921–
    Adventures in celestial mechanics : a first course in the theory
of orbits / by Victor G. Szebehely, Hans Mark. — 2nd ed.
        p.   cm
    Includes bibliographical references.
    ISBN 0-471-13317-5 (cloth : alk. paper)
    1. Orbits.   2. Celestial mechanics.   I. Mark, Hans, 1929–
II. Title.
QB355.S974   1997
521'.3—dc21   97-12770
                                                        CIP

Printed in the United States of America

10 9 8 7 6 5 4 3 2

A TRIBUTE TO VICTOR G. SZEBEHELY

Victor G. Szebehely
1921–1997

It is with a heavy heart that I write these words. On September 13, 1997, my friend, colleague, and mentor, Professor Victor G. Szebehely, died at his home in Austin. I have to confess that it has been very hard for me to carry on with this book without him. He was the guiding spirit of the work.

Victor and I had much in common. We were both refugees from Europe—he from Hungary and I from Austria—fleeing the twin scourges of Nazism and Communism. We both became Americans and we both worked on technical projects related to the national security. We both came to love The University of Texas. Finally, we both developed a strong interest in space exploration, and Victor made important contributions to the success of our journeys to the Moon.

Where Victor was unique was in his deep understanding of celestial mechanics and his ability to apply this knowledge to the solution of practical problems. Victor did not hesitate to tackle the toughest scientific problem in his field which is the subject of the final chapter in this work: The problem of three bodies. He had the intellectual courage to take on the hardest challenges and the intellectual horsepower to make critical contributions of lasting value.

I would be remiss if I did not mention Victor's personal qualities. In addition to being a man of intellect, Victor was also a man of good will who was honored and respected by all who knew him. Perhaps most important for his friends was his impish sense of humor. We both had our offices on the fourth floor of Woolrich Hall, the aerospace building on our campus. I was on one end of the floor and he was on the other. One morning I was complaining to him about something that had gone wrong with our research funding in the Congress. Suddenly, he proposed that we resurrect the Austro-Hungarian Empire on the fourth floor and raise the Imperial banner "with appropriate salutes" every morning. "Maybe," he said, "that will solve your problem!" I laughed and promptly forgot what was upsetting me.

Victor Szebehely was a great man whose influence was widespread. I was one of the people who came into his orbit and I am proud to have been his student. I mourn him and I miss him. Rest in peace, my friend, and go with God.

September 1997   HANS MARK

CONTENTS

PREFACE

This volume is the second edition of *Adventures in Celestial Mechanics* published by Victor G. Szebehely in 1989 at the University of Texas Press. The subject of this edition is the same as the previous one which was to quote the earlier introduction "to study the motion of natural and artificial bodies in space." The work is still intended as a textbook for a first course in orbital mechanics and spacecraft dynamics and we have attempted to produce a second edition that maintains the spirit of the first. This was also stated succinctly in the introduction of the first edition: "fundamental ideas will be emphasized and will not be cluttered up with details that are available in the immense literature of this field."

Having described the similarities between this book and the previous one, we should say a word about the changes. The principal difference between the two editions is that we have added some material that strengthens the treatment of the "artificial bodies in space." A chapter on rocket propulsion has been added that describes what must be done to get things into space. We have included a chapter on elementary spacecraft dynamics so that we discuss not only trajectory maneuvers but also how spacecraft are stabilized and oriented. Finally, we have included a chapter on the exploration of the solar system in which the "natural" and "artificial" bodies are treated together. This area is one of the genuine triumphs of modern science and engineering, and it constitutes the most important modern application of celestial mechanics. Therefore we felt that it was necessary to address it even in an elementary course.

In addition to these major changes, there are minor ones as well. In several instances (e.g., Lambert's theorem and gravity-assist trajectories) we have included details that were not present in the first volume. We have also somewhat expanded the discussion of the three-body problem to include chaotic motion in nonlinear systems.

For the most part, therefore, the second edition is similar to the first. Each of the chapters contains some numerical examples so that students will become familiar with how various calculations are performed. Problems are also included at the end of each chapter. Finally, appropriate references are mentioned at the end of each chapter and also in the appendix.

Many people helped us to write this book. We are grateful to these colleagues in particular: Professor Roger Broucke for his help in developing the derivation of Lambert's theorem, Professors Wallace Fowler and Bob E. Schutz for their help in writing and revising Chapter 9 (Elements of Spacecraft Dynamics), and Professor Raynor L. Duncombe for carefully reading and commenting on the manuscript. We owe a very special debt of gratitude to Ms. Maureen A. Salkin who did a superb job typing the entire manuscript. In addition, Ms. Salkin made important editorial suggestions that significantly improved the quality of the work. Finally, we would like to thank all of the students who were in our classes during the years that we have taught this course at The University of Texas at Austin. These young people provided us with continuing stimulation and inspiration which made it a great pleasure for both of us to work on this project.

VICTOR G. SZEBEHELY
HANS MARK

Austin, Texas
August 1997

CHAPTER 1

ON THE SHOULDERS OF GIANTS: AN HISTORICAL REVIEW

People have looked at the stars since the dawn of history. The obvious "permanence" of the heavens and the regularity of the motions executed by the Sun, the Moon, and the planets soon led people to look for explanations. Each of the major civilizations produced a "cosmology" that was based on more or less crude observations and was melded with the myths of the civilization. These "theories" of the cosmos were important in that they were early attempts to understand how the universe works. While many of these had philosophical and perhaps literary value, they lacked what is essential in a modern scientific theory: predictive value. None of these theories were able to make really accurate predictions of phenomena such as eclipses or were able to explain why the observed regularities in the planetary motions exist.

During the fourth and third centuries before the birth of Christ, there was a great flowering of civilization in Greece. Philosophical schools were established by a number of people, and one of the major topics of interest was cosmology. Many theories were set forth, including at least one that put the Sun at the center of the solar system. Aristarchus of Samos (ca. 270 B.C.) developed some clever techniques for measuring both the sizes and the distances to the Moon and to the Sun. Although his methods were crude, and in the case of the Sun somewhat flawed, he did conclude from his observations that the Sun must be much larger than Earth. It was from this "measurement" that Aristarchus was the first to conclude that the Sun, rather than Earth, should be placed at the center of the solar system. At about the same time, Eratosthenes of Alexandria (ca. 276 B.C.) actually measured the radius of Earth by comparing the length of the shadow cast by similar vertically placed rods, one in Syene and the other at Alexandria, at high noon on the first day of summer. The value he calculated was within 20% of the ones obtained by modern instruments.

However, by far the most influential natural philosopher of the period was Aristotle (384–322 B.C.). He taught that the only way to understand the world was by the application of pure *reason*. This approach led him to two conclusions that were to impede progress for more than 18 centuries. Aristotle argued that it was *common sense* to conclude that Earth is fixed in space and located at the center of the universe. Furthermore, he said that the gods lived in heaven, and thus all motion in the heavens had to be "perfect," by which he meant uniform and circular. Unlike Aristarchus, most philosophers of the day did not attach much value to detailed observations and measurements. Thus, Aristotle's views prevailed because of his enormous influence; he was, after all, the teacher of Alexander the Great.

The cosmology of Aristotle was developed in a systematic way by Claudius Ptolemaeus (ca. A.D. 140). Ptolemaeus was a Greek who lived in Alexandria, where he produced a monumental treatise called the Almagest that included a detailed section on cosmology. He placed Earth at the center of the universe and said that the stars were fixed on a large sphere that rotated around the central Earth once every day. Since the Sun, the Moon, and the planets all moved relative to the stars, they were said to be attached to different spheres, all rotating in uniform motion around Earth. To explain the complex (and sometimes even retrograde) motion of some of the planets, smaller spheres were attached to the larger ones, and the planet was then located on the surface of the small sphere. This sphere also would rotate with uniform angular velocity, thus preserving the Aristotelian doctrine of uniform circular motion for this complex system of spheres upon spheres. Using what were called *cycles* and *epicycles,* this model turned out to be remarkably accurate given the state of astronomical instruments in the second century A.D. While the model of cycles and epicycles had descriptive value, it did not explain why the stars and planets moved the way they do.

It took more than a thousand years to change this stage of affairs. In the thirteenth century, Roger Bacon, an English cleric, was the first to propose that hard knowledge (theories, if you will) must be based on observation and that these observations must be rigorously controlled and objective; that is, they must be repeatable by any observer. What we now call the "modern" science slowly evolved from Bacon's ideas. In a very real sense, Bacon was the one who set the stage for the great scientific achievements of the renaissance period.

It can be argued that the very first important and genuine application of the modern scientific method was the complete and detailed understanding of how the solar system works. All of the hallmarks of how modern science is done are there: the introduction of a new hypothesis, perhaps even for the wrong reason; the development of a reliable body of measurements; the rejection of the existing theory by showing that the measurements support the new hypothesis; and, finally, the demonstration that the new theory can explain things that could not be understood previously. The first tentative steps were taken by Nicolaus Copernicus (1473–1543) (Mikolaj Kopernik in Polish), who introduced the hypothesis of a solar system with the Sun, rather than Earth, at the center. In the first instance, he did this for a practical reason, since it was an attempt to simplify the calculations necessary to maintain an accurate calendar. Using the older, geocentric model of the solar system developed by Claudius Ptolemaeus (Ptolemy), calendar calculations had become very complicated as better measurements became available. Copernicus nursed the hope that, by placing the Sun at the center of the solar system, he could reduce the number of parameters necessary to make good predictions of the celestial phenomena and events that determined the calendar. In this effort, Copernicus was only partially successful. However, what is important is that a "truth" dawned on him during the process of his work, which was that the Sun really is located at the center of the solar system. As a conservative clerical lawyer, Copernicus was shocked by his own hypothesis, and he never published anything that contained the absolute assertion that the Sun was at the center of the solar system during his lifetime. His major work, "De Revolutionibus Orbium Coelestium" was published only after his death. We thus have the accidental stumbling on a major "truth" that occurs so often in the modern scientific process.

A second feature of scientific discovery is accurate and reliable experimentation. Tycho Brahe (1546–1601) was the most important exponent of this process of understanding the solar system. Tycho was a Danish aristocrat who received a cosmopolitan and international education. He took up observational astronomy as a hobby and, because of his great wealth, was able to build what was, for his time, the finest astronomical observatory in the world. It was called the Uranienborg (castle of the sky) and was located on the Island of Hven near Copenhagen. Tycho, for the first time, made accurate measurements of the positions of the Sun, the Moon, the planets, and the stars. What is more important is that he made these observations systematically over more than 20 years. He was therefore the first to produce accurate ephemeris tables, and as we shall see, these eventually turned out to be of decisive importance. Three years before his death, Tycho was forced to leave Denmark. The Emperor Rudolf II then invited Tycho to become the Astronomer to the Imperial Court in Prague. It was there that he met Johannes Kepler, which led finally to the great breakthrough.

Galileo Galilei (1564–1642) also made a most important "experimental," or observational, discovery by being the first person to turn the newly invented telescope toward the sky. By observing that the four large moons of Jupiter execute more or less circular orbits around the planet, he had discovered a small system that demonstrated clearly how the larger solar system works. It was this observational discovery that provided a convincing argument that the Copernican hypothesis regarding the position of the Sun at the center of the solar system was correct. The contributions of both Tycho and Galileo were critical: Galileo's was qualitative, but it gave others the courage to go ahead. Galileo was also an enthusiastic and articulate controversialist and he was able to engage the educated public in the cosmological debate. It is interesting that the great work of Copernicus, "De Revolutionibus Orbium Coelestium" was put on the Index by the Vatican in 1616 (70 years after publication), only after Galileo began his propaganda campaign for the Copernican system. Finally, it was Tycho Brahe who provided the trustworthy numbers.

As important as these contributions were, the real intellectual breakthrough came from Johannes Kepler (1571–1630). Kepler was the son of a German noncommissioned officer. His talents in mathematics were recognized very early in his life, and he was educated by the local clerical authorities. Eventually, he was appointed Professor of Mathematics at the University of Graz in Austria, where he began his astronomical studies. He believed in the heliocentric hypothesis, and he made several attempts to develop a mathematical model of the solar system based on placing the Sun at the center. All of these models failed to fit the observations, and so, in 1599, he decided that he would go to work for the man who had the best observations, Tycho Brahe. Tycho had been exiled from his native Denmark in 1598 and had moved to Prague. Kepler applied for the post as Tycho's assistant at the Imperial Court and his application was accepted. Unfortunately, Tycho died shortly after Kepler arrived in Prague, and Kepler was forced to fight a lengthy legal battle to get access to Tycho's ephemeris tables. Eventually, he succeeded and this is when his great work began.

Perhaps the single most difficult thing that must be done in the process of scientific discovery is to abandon that which was previously taken to be the "truth." Habits of thinking are hard to break, but this is exactly what Kepler did when he abandoned the old Greek idea enshrined by Aristotle that all heavenly bodies must execute perfect motion, meaning that their motion must be in circular orbits moving at uniform speed. In doing his calculations, Kepler could not explain Tycho Brahe's observations of the motion of Mars with the assumption that Mars was moving in a uniform circular orbit around the Sun. It was at this point that Kepler made his great breakthrough. He chose to abandon Aristotle and to believe the observations of Tycho Brahe and turned the question around: Given the observations of Tycho, what kind of orbit does Mars execute? It was in answering this question that Kepler discovered his quantitative laws of planetary motion. A good argument can be made that Kepler's step was actually the most difficult one in the entire process, because he had to do two things that involved great intellectual risks. First, he had to abandon the centuries-old idea of uniform circular motion and, second, he had to believe Tycho's observations to derive his laws. It was the complete rejection of the old and the leap of faith in the new measurements that made Kepler's achievement the most remarkable one in the entire story.

Kepler's laws of planetary motion may be stated as follows:

1. Planets move around the Sun in elliptic orbits with the Sun located at one focus of the ellipse.
2. As the planet moves in its orbit around the Sun, equal areas as measured from the focus are swept out in equal times. (This law implies that the planet moves more rapidly when it is close to the Sun compared to when it is farther away.)
3. The square of the period of the orbit is proportional to the cube of the semimajor axis of the elliptic orbit.

The final chapter in this history came when Isaac Newton realized that Kepler's laws were the consequences of more basic principles, the law of universal gravitation and the so-called second law of motion, which relates the acceleration of an object with the force that is applied to move it. These two principles were sufficient to explain Kepler's laws and much else as well. If Kepler was the one who broke with the past, it was Newton who looked to the future. As Newton put it, "If I have been able to see a little farther, it is because I stood on the shoulders of giants."

Isaac Newton was born at Woolsthorpe in Lincolnshire on Christmas Day in 1642. He died almost 85 years later in 1727. He received his B.A.

degree in Cambridge in 1665. In 1669, when his professor Isaac Barrow resigned, he requested that Newton be given his professorship. Newton's complete dedication to his work resulted in headaches, sleepless nights, irregular eating habits, and finally a nervous fatigue at 50 years of age. He mentions these problems in his notes on the computations of the motion of the Moon.

In 1687, before switching to administrative activities as the Warden and, in 1699, as the Master of the Mint, his book, entitled *Philosophiae Naturalis Principia Mathematica,* was published by the Royal Society of London. It is interesting to see how dynamical problems can become complicated at Newton's insistence that they be solved using geometry instead of calculus. This makes the *Principia* a hard book to read and leads to the question of why Newton, one of the inventors of calculus, did not use calculus in his book. Newton had used calculus to formulate and solve some of the problems presented in the *Principia* but, being afraid of criticism, described his work using geometry.

Newton's conflicts with Leibnitz concerning the discovery of calculus are well represented in the literature, and this may be another reason why geometry dominates the *Principia.* Their controversy regarding the deterministic nature of dynamics and celestial mechanics is less known. Today, Newtonian mechanics is sometimes erroneously associated with complete predictability in dynamics, which was Leibnitz's dogma and was not accepted by Newton. At this point, Laplace's demon enters the picture: knowing all initial conditions and all laws of nature and predicting the future. Laplace takes the side of Leibnitz. (See the list of bibliography at the end of this chapter.)

In 1665, because of plague, Newton left Cambridge and went back to his birthplace, where he could work undisturbed. The unverified apple incident, which could have happened here, describes the importance of connecting seemingly unrelated phenomena; in this case, falling stones (or apples) on the one hand and planetary motion on the other. In fact, Newton describes the idea leading to artificial satellites with the following thought experiment: If stones are thrown from the top of a mountain with small horizontal velocities, they will hit the ground, but as the velocity is increased, circular and elliptic orbits are obtained around Earth. It was here, amid conditions of creativity, concentration, and peace, that Newton arrived at the general theory of gravitation.

Newton became the president of the Royal Society at the age of 60 and was knighted by Queen Anne in 1705. He died in 1727 and is buried at Westminster Abbey in London.

Since, in this book, we wish to concentrate on dynamics and celestial mechanics, for a description of Newton's many other significant scientific contributions (e.g., his *Opticks,* published in 1704), the reader is referred to the literature.

Since Newton's laws of dynamics and his law of gravitation will be described here, a few general historical comments might be appropriate.

Newton's three laws of motion, forming the basis of dynamics, are as follows:

1. Every body perseveres in its state of rest or uniform straight-line motion unless it is compelled by some impressed force to change that state.
2. The change of motion is proportional to the motive force impressed and takes place in the same direction as the force.
3. Action is always contrary and equal to reaction.

There are many variations of these laws, some by Newton himself, who made changes and corrections. Also, differences exist in the literature as the laws were translated from the original Latin text. Once again, the soundest language, mathematics, comes to our aid. Using the concept of linear momentum (which Newton called motion), we can express the first and second laws by the equation

$$\frac{d(m\mathbf{v})}{dt} = \mathbf{F} \tag{1.1}$$

Note that Newton did not mention acceleration when giving his laws of motion. For a constant value of the mass, the above equation should read: $m(d\mathbf{v}/dt) = \mathbf{F}$. Our textbooks use the concept of acceleration and give Newton's law as $m\mathbf{a} = \mathbf{F}$. This is of less generality than Newton's original formulation, which is applicable to variable mass and, therefore, for rocket propulsion.

Newton's law of gravitation, as discussed in his *Principia,* was mentioned before. The gravitational force acting between two bodies of mass m and M is proportional to the product of the masses and inversely proportional to the square of the distance between them. In vector form

$$\mathbf{F} = \frac{GmM}{|\mathbf{r}|^3}\,\mathbf{r} = G\frac{mM}{r^2}\,\hat{r} \tag{1.2}$$

where $\hat{r}$ is the unit vector pointing in the direction $\mathbf{r}$ and G is the gravitational constant that determines the "strength" of the gravitational field.

Probably nothing describes Newton better than one of his own statements: "I seem to have been only like a boy playing on the seashore and diverting myself now and then finding a smoother pebble or a prettier shell than ordinary, while the great ocean of truth lay all undiscovered before me."

It is a common error to believe that the behavior of the solar system and the rules of orbital mechanics were completely understood as a result of the work of Isaac Newton. He took a giant step, but many critical questions remained unanswered. Newton solved what we call the "problem of two bodies," which means that he developed the means to predict the motion of two bodies interacting through the gravitational field. For a system of more than two bodies, Newton's equations cannot be solved. Fortunately, the solar system is dominated by the Sun, which accounts for more than 99.8% of its entire mass. Thus, to a very good approximation, the motion of each planet can be calculated *as if* only the Sun and that planet counted. Thus, Newton was able to deduce his laws. With the advent of more accurate astronomical measurements in the eighteenth century, discrepancies appeared that could only be explained by taking into account the effects of the other planets in the solar system.

Following Newton's work, several brilliant astronomers and mathematicians used Newton's laws and methods to attack a number of important problems. The first of these was Edmund Halley (1656–1742), who observed and calculated the orbit of the comet named after him using Newton's laws of motion. Studying several cometary orbits, he established the facts that, contrary to planetary orbits, some comets had large angles of inclination and that some had periodic orbits. Halley's contributions were numerous and important to celestial mechanics, but his insistence on and support of the publication of Newton's *Principia* probably represent his greatest influence on today's celestial mechanics.

The Swiss-born mathematician Leonhard Euler (1707–1783) was a student of Johann Bernoulli. In 1727 Euler went to St. Petersburg in Russia for 14 years and was associated there with the Imperial Academy. From there, at the invitation of Frederick the Great, he went to Berlin, where he remained for 25 years. He returned to St. Petersburg at the invitation of the czarina, Catherine the Great, in 1766.

Euler's work on the motion of the Moon was of considerable interest to Catherine the Great as his lunar tables and his second lunar theory, published in 1772 under the title *Theoria Motuum Lunae* in the Communications of Petropolis, helped the navigation of ships in the Russian Navy. Before it appeared in its published form, his lunar theory was used by the Astronomer Royal, Nevil Maskelyne, in the British Nautical Almanac as the basis for the lunar ephemeris. These tables were first published in 1767 and were used by the British Navy for navigation. (These were probably the first, but certainly not the last, uses of celestial mechanics by the military.)

Newton's most important successors, who truly extended his methods, were two Frenchmen whose lives spanned the last years of the eighteenth century and the first years of the nineteenth: Joseph Louis Lagrange (1736–1813) and Pierre Simon de Laplace (1749–1827). Lagrange was born in Turin, Italy, where he was appointed professor of geometry at the artillery academy at the age of 19. In 1766, he went to Berlin, filling Euler's vacated position at the invitation of Frederick the Great, where he spent 20 years. The next invitation came from Louis XVI to Paris, where he became professor at the École Polytechnique in 1797. His apartment in Paris was in the Louvre; he was buried in the Pantheon.

Lagrange's announcement concerning the triangular libration points in the Sun–Jupiter system and his prediction of the possible existence of asteroids in these regions date from 1772. Observational astronomers did not verify the existence of these bodies for another 134 years. In this case, theory was certainly ahead of observation. His work on the solar system using the method of variation of parameters (1782) is one of the fundamental contributions in celestial mechanics.

Lagrange's celebrated *Mécanique Analytique* was published in 1788.

Laplace was born in Beaumont-en-Auge and became professor at the École Militaire in Paris at the age of 18. One of his major contributions concerned the stability of the solar system (1773, 1784), for which he developed the methods of perturbation theory to solve the many-body problem. After a lengthy series of calculations, he concluded that the solar system was indeed stable and that Newton's famous "clockwork universe" really existed. As things turned out, Laplace was wrong, and the problem of "stability" is still unsolved. Laplace also introduced the concept of the potential function and what is known today as Laplace's equation (1785). His lunar theory, published in 1802, followed Euler's. The five volumes of his *Mécanique Céleste* were published between 1799 and 1825.

Although the perturbation methods introduced by Laplace did not yield an answer to the stability question, they were extremely useful in making more accurate calculations of the behavior of planets, comets, and asteroids. The most spectacular application of perturbation theory was the discovery of the eighth planet, Neptune, because of the small perturbations the planet causes in the motion of the planet Uranus. John Couch Adams and U. J. J. Leverrier performed these calculations in 1845 and predicted the position of Neptune. In the next year, J. F. Encke and H. L. d'Arrest found Neptune essentially where it was supposed to be. In the early years of this century, Percival Lowell and William H. Pickering tried to do the same thing by looking at small perturbations in the orbit of Neptune. The theoretical work done by Lowell and Pickering between 1910 and 1917 was detailed and extensive. Lowell died in 1917, but Pickering continued to work on the problem. Eventually, another search for a trans-Neptune planet was initiated, and in 1930, the young astronomer Clyde W. Tombaugh discovered Pluto. The "predictions" of Lowell and Pickering could not have had anything to do with the discovery of Pluto since the planet turned out to be much too small to affect Neptune in the way Lowell and Pickering had calculated. In any event, these remarkable achievements effectively completed the inventory of planets in our solar system. They were stimulated by the development of perturbation theory.

The most important contributor to celestial mechanics in the final years of the nineteenth century and the early years of the twentieth was another Frenchman, Henri Poincaré (1854–1912). He was one of the most prolific writers in the field of mathematics and celestial mechanics, contributing more than 30 books and 500 memoirs. The three volumes of his *Méthodes Nouvelles de la Mécanique Céleste* appeared in 1892, 1893, and 1899 and have been recently translated into English by NASA. This was followed by his *Léçons de Mécanique Céleste* in 1905–1910. Concentrating on the problem of three bodies, Poincaré established the concept of nonintegrable dynamical systems. His theorem seriously affected the results of workers who intended to show the stability of the solar system by representing the orbital elements of the planets in Fourier series. Since these series, in general, are conditionally convergent or divergent according to Poincaré's theorem, the "solutions" do not show stability. Thus Laplace's conclusion of a century earlier was shown to be wrong. Poincaré's work also provided the first instance of what is now called "deterministic chaos." The problem of three bodies is described by a complete set of deterministic equations. Yet, the behavior of the three-body system may become "chaotic," which in this case means unpredictable, under certain conditions. It may very well be that this will turn out to be Poincaré's lasting legacy.

In recent years, a most significant development has furthered the science and engineering of orbital operations and that is the advent of artificial satellites and spacecraft. The demands of space navigation have clearly been a major factor in the recent progress of celestial mechanics. This effort has been greatly enhanced by the advent of high-performance digital computers, which make the approximation methods mentioned

earlier less necessary. The truly fabulous accuracy of spacecraft navigation would not be possible without high-speed digital computers. For example, to put the *Pioneer 11* spacecraft into the correct trajectory around
Jupiter so that it would fly past Saturn some years later required a naviga-
tional accuracy of better than one part in 10 million.

Finally, there are some very important scientific questions that are still
open. Is the solar system ultimately stable? This question has not been answered in a rigorous mathematical sense. Once again, numerical methods
are critical to research this question. Related to the question of stability is
that of chaotic motion. Can the "Earth crossing" asteroids be explained
using the principles of chaos theory? Thus, orbital and celestial mechanics, even though it is the oldest field in "modern science," still presents
problems that are at the very frontier of knowledge.

What is clear is that celestial mechanics is a living field and more research is certain to reveal important and even startling new results.

The reader interested in historical details will enjoy some of the books
listed in the Appendix: Andrade (1954); Bate, Mueller, and White (1971);
Beer and Strand (1975); Koestler (1959); and Lerner (1973). In addition,
Men of Mathematics, by E. T. Bell, Simon & Schuster, New York (1937);
The Great Ideas Today, edited by R. M. Hutchins and M. J. Adler, Ency-
clopaedia Britannica, Inc. (1973); *From Galileo to Newton,* by A. R. Hall,
Dover, New York (1981); and *The Space Station,* by H. Mark, Duke University Press, Durham, North Carolina (1987), are recommended. Regarding nondeterministic dynamics and uncertainties in celestial mechanics, see J. Lighthill's "The Recently Recognized Failure of Predictability
in Newtonian Dynamics," *Proceedings of the Royal Society,* Vol. A407,
pp. 35–50, 1986, and I. Prigogine's (1980) book listed in the Appendix.

For additional fascinating details of the early history, see "Copernicus
and Tycho," by O. Gingerich, *Scientific American,* Vol. 229, No. 6, pp.
86–101, 1973. For Newton's contributions to cosmology, see *The First
Three Minutes,* by S. Weinberg, Bantam Books, New York (1977).

CHRONOLOGICAL LIST OF MAJOR CONTRIBUTORS
TO CELESTIAL MECHANICS

Aristotle,	384–322 B.C.	I. Newton	1642–1727
C. Ptolemaeus	100–178	G. W. Leibnitz	1646–1716
N. Copernicus	1473–1543	E. Halley	1656–1742
T. Brahe	1546–1601	L. Euler	1707–1783
G. Galilei	1564–1642	A. C. Clairaut	1713–1765
J. Kepler	1571–1630	J. D'Alembert	1717–1783
R. Descartes	1596–1650	J. H. Lambert	1728–1777

J. L. Lagrange	1736–1813	P. H. Cowell	1870–1949
W. F. Herschel	1738–1822	W. De Sitter	1872–1934
J. E. Bode	1747–1826	F. R. Moulton	1872–1952
P. S. Laplace	1749–1827	T. Levi-Civita	1873–1941
M. Legendre	1752–1833	K. F. Sundman	1873–1949
K. F. Gauss	1777–1855	E. T. Whittaker	1873–1956
S. D. Poisson	1781–1840	H. C. Plummer	1875–1946
J. F. Encke	1791–1865	W. Hohmann	1880–1945
G. G. DeCoriolis	1792–1843	G. A. Shook	1882–1954
J. F. W. Herschel	1792–1871	G. D. Birkhoff	1884–1944
P. A. Hansen	1795–1874	W. M. Smart	1889–1975
K. G. J. Jacobi	1804–1851	G. E. Lemaitre	1894–1966
W. R. Hamilton	1805–1865	C. L. Siegel	1896–1981
U. J. J. Leverrier	1811–1877	Y. Hagihara	1897–1979
C. E. Delauney	1816–1872	N. D. Moiseev	1902–1955
J. C. Adams	1819–1892	D. Brouwer	1902–1966
D. Airy	1835–1981	W. J. Eckert	1902–1971
S. Newcomb	1835–1909	A. Wintner	1903–1958
T. N. Thiele	1838–1910	A. N. Kolmogorov	1903–1987
G. W. Hill	1838–1914	G. N. Duboshin	1904–1986
F. F. Tisserand	1845–1896	G. P. Kuiper	1905–1973
H. Bruns	1848–1919	G. M. Clemence	1908–1974
G. H. Darwin	1845–1912	P. Herget	1908–1981
J. H. Poincaré	1854–1912	E. L. Stiefel	1909–1987
C. V. L. Charlier	1862–1934	S. Herrick	1911–1974
P. Painleve	1863–1933	G. A. Chebotarev	1913–1975
E. W. Brown	1866–1938	H. Pollard	1919–1985
C. Burrau	1867–1944	G. Colombo	1920–1984
E. Stromgren	1870–1947		

CHAPTER 2

CIRCULAR ORBITS

In the preceding chapter, in equations (1.1) and (1.2), we defined the laws of motion first developed by Isaac Newton and his universal law of gravitation. It is the combining of these two laws that permits us to calculate the orbit of one body moving around another one under the influence of the gravitational interaction between the two bodies. A particularly simple case to treat is that of circular orbits. We shall assume for the moment that circular orbits are both possible and stable in the gravitational field defined by equation (1.2). This proposition will be proven in subsequent chapters.

The law of gravitation as shown in equation (1.2) is given as

$$\mathbf{F}_G = -\frac{GMm}{r^2}\,\hat{r} \tag{2.1}$$

where $\mathbf{F}_G$ is the force of gravity between the masses M and m. The unit vector $\hat{r}$ points in the direction of the line joining the masses, and r is the distance between the masses m and M. The situation is illustrated in Figure 2.1. For the time being, we shall assume that the masses m and M are point masses. We shall show shortly that for spherically symmetric objects the gravitational field external to the object acts as if the mass were concentrated at the geometric center of the object. The constant G is called the *gravitational constant,* and it determines the strength of the gravitational field. In Figure 2.1, we have assumed that the mass M is lo-

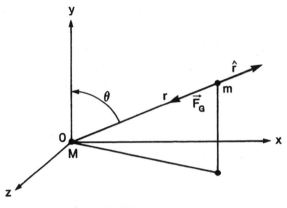

FIGURE 2.1

cated at the origin O of the coordinate system and that it is fixed in space. (We shall soon show that this is equivalent to saying that M is very much larger than m.) Note that the first $\mathbf{F}_G$ points toward the origin, where mass M is located. This happens because the gravitational force is always attractive. Note that the convention of polar coordinate systems requires that the unit vector $\hat{r}$ always points away from the origin. This accounts for the negative sign on the right side of equation (2.1), because $\mathbf{F}_G$ and $\hat{r}$ always point in opposite directions.

Figure 2.2 shows the circular orbit that we have assumed is possible in this case.

We assume that the radius of the circular orbit is R and that the vector

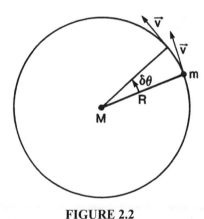

FIGURE 2.2

v is the velocity of the mass m as it moves around the mass m in the circular orbit. There are two forces acting on the mass m: the gravitational force, which points toward the mass M, and the centrifugal force experienced by an object traveling in a circular orbit. If the masses M and m were connected by a string, then the tension in the string would replace the gravitational force and would also be balanced by the centrifugal force.

The centrifugal force can now be calculated using equation (1.1) of the previous chapter:

$$\mathbf{F}_c = m\frac{d\mathbf{v}}{dt} \qquad (2.2)$$

We now need to evaluate the rate of change of the velocity ($d\mathbf{v}/dt$) that appears in equation (2.2). To do that, we shall look at what happens to the orbital velocity vector. Since the gravitational force defined in equation (2.1) on the mass m is constant and since the radius of the circle, R, does not change as the mass m moves in its orbit, the magnitude of the vector **v**, $|\mathbf{v}|$, must also be constant. The rate of change of the velocity vector is therefore determined only by the change in direction as m moves around the orbit, as shown in Figure 2.3. If we consider only small angles, $\delta\theta$, we can look at the way the vector **v** behaves by looking at Figure 2.2. The vector $\Delta\mathbf{v}$ is the change in direction of the velocity vector **v**. Note that this vector, $\Delta\mathbf{v}$, always points toward the mass M at the origin of the coordinate system. Thus,

$$\mathbf{F}_c = m\,\frac{d\mathbf{v}}{dt} = -m|\mathbf{v}|\frac{d\theta}{dt}\,\hat{r} \qquad (2.3)$$

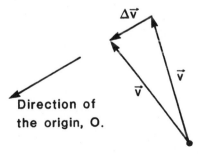

Direction of the origin, O.

FIGURE 2.3

where the differential change in velocity is given by

$$d\mathbf{v} = -|\mathbf{v}|\, d\theta\, \hat{r} \tag{2.4}$$

The *angular velocity* ω of the mass m is defined as $d\theta/dt$ so that the centrifugal force can be rewritten as

$$\mathbf{F}_c = -m\,|\mathbf{v}|\,\omega\,\hat{r} \tag{2.5}$$

and simply writing v for $|\mathbf{v}|$ and recognizing that for a circular orbit of radius R the velocity v is

$$v = R\omega \tag{2.6}$$

we have, for the centrifugal force,

$$\mathbf{F}_c = -\frac{mv^2}{R}\,\hat{r} \tag{2.7}$$

Equating (2.7) to (2.1), we have

$$\mathbf{F}_G = \mathbf{F}_c \qquad G\frac{Mm}{R^2} = \frac{mv^2}{R} \tag{2.8}$$

Equation (2.8) allows determination of the orbital speed v (which is the magnitude) of the velocity vector $\mathbf{v}$ as

$$v = \sqrt{\frac{GM}{R}} \tag{2.9}$$

Note that the mass m appears on both sides of equation (2.8) so that the orbital speed is a function only of the radius of the circle and the magnitude of the mass M. Equation (2.9) can be rewritten in terms of the angular velocity defined in equation (2.6):

$$R\omega = \sqrt{\frac{GM}{R}} \tag{2.10}$$

and so we have

$$R^3\omega^2 = GM \tag{2.11}$$

The angular velocity can be related to the period of the orbit, that is, the time it takes to execute one orbit, by returning to the definition of the angular velocity,

$$\omega = \frac{d\theta}{dt} \quad \text{or} \quad \omega\, dt = d\theta \qquad (2.12)$$

and integrating around one orbit,

$$\int_0^T \omega\, dt = \int_0^{2\pi} d\theta$$

we obtain

$$\omega T = 2\pi \qquad (2.13)$$

where T is defined as the orbital period. Substituting equation (2.13) into (2.11) yields

$$R^3 = T^2 \left(\frac{GM}{4\pi^2} \right) \qquad (2.14)$$

This statement is the third law of planetary motion as stated by Kepler (see Chapter 1) for the special case of circular orbits. It is obvious that the second law is also fulfilled for circular orbits since the orbital speed v is constant so that equal areas are swept out in equal time. In subsequent chapters, we shall show that these statements are valid for elliptic orbits as well.

Equation (2.11) is a very good approximation to the exact relation when we consider the motion of a satellite around Earth in a circular orbit. The approximate result assumes that the mass of the satellite can be neglected when compared to the mass of the central body. The derivation of the exact relation for circular motion utilizes Figure 2.4.

The satellite and Earth are moving around the center of mass of the Earth–satellite system. Since the mass of Earth is always many orders of magnitude larger than the mass of the satellite, the center of mass of the system is at the center of Earth for all practical purposes. As another interesting example, consider a binary star or a binary asteroid where two stars or two asteroids with comparable masses are revolving around each other. The distances from the center of mass are r_1 and r_2; the masses are

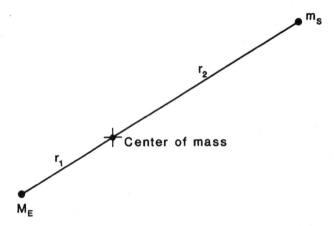

FIGURE 2.4 Motion of a satellite around the center of mass of the system.

m_1 and m_2. In our original problem, $m_1 = M_E$ and $m_2 = m_S$. The forces acting on the satellite are balanced if

$$m_s\omega^2 r_2 = G\frac{M_E m_S}{r^2} \tag{2.15}$$

where $r = r_1 + r_2$ since that is the total distance between the two interacting bodies.

The corresponding equation for Earth is

$$M_E\omega^2 r_1 = G\frac{M_E m_S}{r^2} \tag{2.16}$$

In the first equation m_S and in the second equation M_E are canceled. Since the center of mass is fixed in the system, we have

$$m_S r_2 = M_E r_1 \tag{2.17}$$

Adding the two previous equations (2.16) and (2.17), we have

$$(r_1 + r_2)\omega^2 = G\frac{M_E m_S}{r^2} \tag{2.18}$$

Equation (2.18) may be written as

$$\omega^2 r^3 = G(M_E + m_S) \tag{2.19}$$

This is the exact form of Kepler's law for circular orbits, and we see that the previously obtained equation (2.11) needs a modification whereby instead of the mass of Earth we have the sum of the mass of Earth and the mass of the satellite on the right-hand side. Clearly, in satellite dynamics this makes no difference, and consequently equation (2.11) is correct. On the other hand, if we study bodies revolving around each other with comparable masses, equation (2.19) should be used.

To evaluate the approximation, Kepler's law may be written as

$$\omega^2 r^3 = GM_E \left(1 + \frac{m_S}{M_E} \right) \tag{2.20}$$

If the mass of the satellite is 1 ton, the ratio m_S/M_E becomes of the order of 10^{-22}, which is the error made when the m_S/M_E term is neglected.

An Earth-orbiting satellite usually orbits at an altitude that is substantially smaller than the radius of Earth. This situation is illustrated in Figure 2.5. For a typical space shuttle mission, for example, the orbital altitude h might be 480 km compared to Earth's radius of 6370 km. Under these conditions, it is not obvious that the mass of Earth can be considered as the point mass M that we have used to derive the properties of circular orbits. What we must show is that for an approximately spherically symmetric body like Earth the mass of Earth can be considered as a point mass located at the center of Earth.

To prove this theorem, we shall first introduce the concept of the potential function. The potential function is related to the "potential energy" that an object has in a gravitational field. Consider the situation shown in

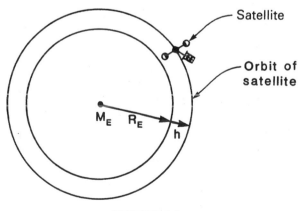

FIGURE 2.5

Figure 2.6. What is the work required to move mass m from r_1 to r_2? The definition of work is

$$W = \int_{r_1}^{r_2} \mathbf{F} \cdot d\,\mathbf{r} \qquad (2.21)$$

Now substituting equation (2.1) for the force of gravity yields

$$W = \int_{r_1}^{r_2} \left[-\frac{GmM}{r^2} \right] dr = GmM \left[\frac{1}{r} \right]_{r_1}^{r_2}$$

$$= GmM \left(\frac{1}{r_2} - \frac{1}{r_1} \right) \qquad (2.22)$$

Note that the work is always negative when pulling away (increasing r) meaning that work must be done and positive when falling inward. From (2.22), it follows that one can associate with each point in the gravitational field a potential energy V_g:

$$V_g = \frac{GmM}{r} \qquad (2.23)$$

so that the work done to move the mass m from r_1 to r_2 is

$$W = V_g(r_2) - V_g(r_1)$$

the difference between the potential functions evaluated at the points r_1 and r_2. The potential function is extremely useful because it is a scalar rather than a vector. The gravitational force $\mathbf{F}_g$ (vector) is always the gradient of the potential function:

$$\mathbf{F}_g = \mathbf{grad}\ V_g \qquad (2.24)$$

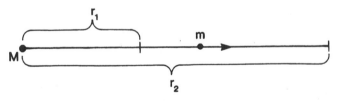

FIGURE 2.6

This relationship is very useful, and we will often see it in subsequent chapters.

The method for proving the theorem that for a uniform sphere the mass can be considered as located at the center is rather lengthy, so we shall only outline the procedure without performing the integrations. The situation is illustrated in Figure 2.7. Earth has been divided into a large number of thin spherical shells, each of which is very thin compared to the radius of the shell. The contribution to the gravitational potential at the location of the mass m due to the mass element dM of the shell is

$$dV_g = \frac{Gm \, dM}{r_1}$$ (2.25)

It is at this point that the usefulness of the potential function becomes apparent. Equation (2.25) is a scalar equation. Had we employed the force equation rather than the potential equation, we would have had to deal with a vector integration. In general, such operations are very much more complex than the scalar integration defined in equation (2.25). When this integration is performed, the result is

$$V_g(\text{shell}) = \frac{GmM \, (\text{shell})}{r}$$ (2.26)

The potential function of the shell, $V_g(\text{shell})$, is therefore identical to what the function would be if the mass of the shell were concentrated at the point O. Since the sphere consists of a concentric series of shells, the

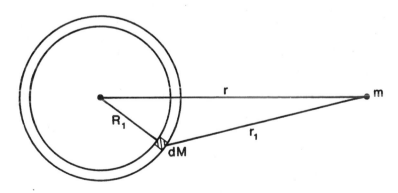

FIGURE 2.7

final integration simply amounts to adding up the gravitational potentials contributed by each shell. Since each shell acts as if its mass is located at point O, the whole sphere must act in the same way, so that

$$V_g(\text{sphere}) = \frac{GmM(\text{sphere})}{r} \qquad (2.27)$$

which proves the theorem.

Having defined the potential energy function, it is useful to use this concept to define the total energy of an object moving in a circular orbit. The kinetic energy is given as

$$E_K = \tfrac{1}{2}mv^2 \qquad (2.28)$$

and this quantity is constant since the orbital speed is constant. The potential energy is defined in equation (2.23). If we start, for example, with a satellite on the surface of Earth and we want to place it in an orbit with a radius r measured from the center of Earth, then the work (energy) that must be expended to get the satellite of mass m to the point r is

$$W = GmM_E\left(\frac{1}{R_E} - \frac{1}{r}\right) \qquad (2.29)$$

where M_E is the mass of Earth (5.98×10^{24} kgm) and R_E is the radius of Earth (6371 km). Note that W is always positive. The total energy of the satellite is then the sum of the kinetic energy and the potential energy:

$$E_r = \tfrac{1}{2}mv^2 + \frac{GmM_E}{R_E} - \frac{GmM_E}{r} \qquad (2.30)$$

The second term on the right is a constant that represents the potential energy of the object when it is located on the surface of Earth. Since the zero point of the energy scale is arbitrary, we can take it as being relative to the surface of Earth. Thus, the total energy can be rewritten as

$$E_T = E_k' + \frac{GmM_E}{R_E} \qquad (2.31)$$

and, therefore, relative to the surface of Earth,

$$E_T = \tfrac{1}{2}mv^2 - \frac{GmM_E}{r} \qquad (2.32)$$

This is a very general relationship that shall be shown to hold for all orbits. For the circular orbits we have been considering, we can use equation (2.9) for the orbital speed of the satellite at radius r:

$$v^2 = \frac{GmM_E}{r} \tag{2.33}$$

so that

$$E_T = \frac{1}{2}\frac{GmM_E}{r} - \frac{GmM_E}{r} \tag{2.34}$$

$$E_T = -\frac{1}{2}\frac{GmM_E}{r} \tag{2.35}$$

Note that the total energy depends only on the radius r and that it is always negative under the conditions described. We shall show in subsequent chapters that equation (2.32) is a general relation valid for any orbit and that equation (2.35) also is valid for any orbit with some modification of the definition of the radius.

It might be surprising to find that a number of practically important problems can be treated using the simple results of Chapter 2. Some of these are discussed in the following examples.

Introductory chapters from Danby's (1962), McCuskey's (1963), and Ryabov's (1959) books are recommended (see references listed in the Appendix).

EXAMPLES

2.1. Consider a satellite on a circular orbit at an altitude of 100 km (=100,000 m = 328,100 ft = 62.137 statute miles). For the computation of the circular speed of the satellite, the values of the constants are given in Appendix 2. Equation (2.10) gives 7.844 km/s, or 17,548 mph, for the circular velocity. The satellite is moving approximately 33 times as fast as a commercial jet aircraft. In this result, no drag effects due to Earth's atmosphere are included and Earth is assumed to be a homogeneous spherical body. The corrections caused by these effects are small for a few revolutions and will be discussed later. The "mean motion" using $\omega = v_c/r$ becomes 1.211×10^{-3} rad/s = 4.163

deg/min, and the period $T = 2\pi/\omega = 5188.43$ s $= 1$ h 26 min 28.4 s. (Note that the angular velocity ω is sometimes called the *mean motion* in orbital mechanics. We will use this term on occasion.)

   If the altitude is 500 km, the above results change little. In fact, the circular speed $v_c = 7.613$ km/s, the mean motion $\omega = 1.107 \times 10^{-3}$ rad/s, and the period P = 1 h 34 min 34 s.

2.2. The second example treats the problem of a 24-h (geostationary or geosynchronous) satellite: Equation (2.9) might be slightly modified by realizing that usually the altitude of a satellite above the surface of Earth is given, rather than its distance from the center of Earth. This relation is

$$r = R_E + h \tag{2.36}$$

where R_E is the radius of Earth and h is the altitude of the satellite above Earth. So equation (2.9) becomes

$$v_c = \sqrt{\frac{GM_E}{R_E + h}} \tag{2.37}$$

This satellite has a period of 24 h; consequently, if its orbital plane is in Earth's equatorial plane, then it will revolve around Earth so that it will always be above the same point of Earth. This satellite is used for communication purposes. From the above period, the mean motion can be computed. From the mean motion, the circular speed as well as the height or the altitude of the satellite can be obtained. Since the period $T = 24$ h $= 86,400$ s, the mean motion becomes

$$\omega = \frac{2\pi}{T} = 7.27 \times 10^{-5} \text{ rad/s}$$

The period and the altitude are connected by

$$T = \frac{2\pi}{\sqrt{GM_E}}(R_E + h)^{3/2} \tag{2.38}$$

from which the altitude becomes

$$h = \left(\frac{T}{2\pi}\right)^{2/3}(GM_E)^{1/3} - R_E = 35,863 \text{ km}$$

or 22,280 miles. The above equation follows from the definition of the period of the orbit and the relationship $r = R_E + h$.

Another approach is to use the relation between the mean motion and the altitude:

$$\omega^2(R_E + h^3) = GM_E$$

from which the altitude may be computed as

$$h = \left(\frac{GM_E}{\omega^2} \right)^{1/3} - R_E$$

which is identical to the previous result obtained above. The circular velocity of the satellite is given by $v_c = \omega(h + R_E)$. This gives 3.072 km/s = 11,058 km/h, or 6871 mph.

The triangles of Figure 2.8 show the relation between the circular velocity of the satellite, which will now be denoted by v_{cs}, and the circular velocity of a point on the equator, v_{cE}, as

$$\frac{v_{cs}}{v_{cE}} = \frac{h + R_E}{R_E} = \frac{h}{R_E} + 1$$

The circular velocity of the equator is obtained from

$$v_{cE} = 2\pi \frac{R_E}{T_E}$$

where $T_E = 24$ h. This gives $v_{cE} = 1669$ km/h, or 1038 mph. This allows us an alternate way to compute the circular velocity of a synchronous satellite using the previously obtained value for the altitude:

$$v_{cs} = \left(\frac{h}{R_E} + 1 \right) v_{cE}$$

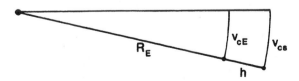

FIGURE 2.8 Geosynchronous satellite.

2.3. This example is related to the approximately circular lunar orbit. The period is approximately 27.32 days, from which the mean motion, the altitude above Earth, and the velocity might be computed. The mean motion is given by

$$\omega_M = \frac{2\pi}{T_M}$$

where T_M is the lunar period. This gives $\omega_M = 2.66 \times 10^{-6}$ rad/s = 13.18 deg/day.

The distance between the center of Earth and the center of the Moon, r, is related to the mean motion by

$$\omega_M^2 r^3 = G(M_E + m_M)$$

which gives $r = 384{,}400$ km.

The circular speed of the Moon may be obtained from $v_c = \omega_M r$, which gives 3594 km/h, or 2233 mph. An easy-to-remember approximate value is 1 km/s.

2.4. The fourth example will be the preparation of a plot showing the periods (T in hours) of satellites in circular orbits around Earth versus their altitude h up to 1000 km. The basic relation is given by equation (2.38).

When the constants are substituted, we have

$$T = 1.4082\left(1 + \frac{h \, (\text{km})}{6378.12}\right)^{3/2}$$

Figure (2.9) shows the plot of the $T(h)$ relation. This is approximately a straight line, which suggests the next example.

2.5. This example will show the derivation of a simple, approximately linear relation between the period and the altitude for Earth satellites in circular orbits. Note that the exact relation is given by equation (2.38), from which we may obtain

$$T = \frac{2\pi}{\sqrt{GM_E}} \, R_E^{3/2}\left(1 + \frac{h}{R_E}\right)^{3/2}$$

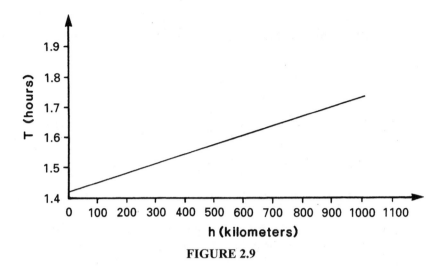

FIGURE 2.9

The factor in front of the parenthesis represents the period a satellite with zero altitude would have, $T_0 = 1.40815$ h. The equation for the period might be written as

$$T = T_0\left(1 + \frac{h}{R_E}\right)^{3/2}$$

The binomial expansion of the factor of T_0 is

$$\left(1 + \frac{h}{R_E}\right)^{3/2} = 1 + \frac{3}{2}\frac{h}{R_E} + \frac{3}{8}\left(\frac{h}{R_E}\right)^2 + \cdots$$

Therefore, the linear relation between altitude and period becomes

$$T_L = T_0\left(1 + \frac{3}{2}\frac{h}{R_E}\right)$$

The error between the exact and the approximate equations is $T - T_L$, and the relative error is $(T - T_L)/T$.

If this error is to be less than 1%, we have

$$\frac{T - T_L}{T} \le 0.01$$

In this equation, T and T_L depend on the ratio h/R_E, which we shall denote by x. In this way we have

$$\frac{(1+x)^{3/2} - (1 + \frac{3}{2}x)}{(1+x)^{3/2}} \le 0.01$$

This inequality leads to a cubic equation for x that may be written as

$$x^3 + 0.7043x^2 - 0.0609x - 0.0203 \le 0.01$$

The only positive root is $x_0 = 0.1887$, which means that if $x = h/R_E$ is less than the above value given for x_0, the error of using the linear equation instead of the exact equation for the period will be less than 1%. The range of applicability of the linear solution is from $h = 0$ to $h = 0.1887R_E$, or

$$0 \le h \le 1203 \text{ km}$$

Note that this problem can be solved by either iteration of the inequality

$$0.99\,(1+x)^{3/2} \le 1 + \frac{3}{2}x$$

or by using the well-known formula for the solution of cubic equations.

2.6. Newton computed the ratio of Earth's gravitational force and the centrifugal force at the equator and obtained "approximately 300." This example is aimed at computing this ratio with the constants given in Appendix 2.

The centrifugal force (per unit mass) is

$$F_c = \omega^2 R_E$$

where R_E is Earth's equatorial radius and ω is the angular velocity of Earth's rotation.

The gravitational force per unit mass is

$$F_g = \frac{GM_E}{R_E^2}$$

The ratio is

$$\frac{F_g}{F_c} = \frac{GM_E}{R_E^3 \omega^2} = 288.3$$

2.7. To compute the mean density of the planets and of the Sun, the values given for their radii and masses in Appendix can be used. Substituting into the formula for density,

$$\rho = \frac{\text{mass}}{\text{volume}} = \frac{3M}{4\pi R^3}$$

and remembering that the density of water is 1 g/cm³, the values computed will show the ratio of the density of the planets to that of water. The mean density of the Sun is 1.41 g/cm³. For the planets, the values are as follows:

Planet	Mercury	Venus	Earth	Mars	Jupiter	Saturn	Uranus	Neptune	Pluto
Density (g/cm³)	5.13+	4.97	5.52	3.94	1.33	0.69	1.56	2.27	4?
Velocity (km/s)	47.8	35	29.8	24.2	13.1	9.7	6.8	5.4	4.7?

2.8. The orbital speeds of the planets are computed from the formula

$$v = \left[\frac{G(M_S + M_P)}{a_P} \right]^{1/2}$$

where M_S and M_P are the masses of the Sun and the planet and a_P is the mean distance of the planet. The values are given in the table above.

2.9. Compare the compactness of the solar system with that of the hydrogen atom. The compactness or denseness can be measured by the ratio of the distance a of a planet (electron) from the Sun (nucleus) to the radius R_S of the Sun (nucleus). For the hydrogen atom, we have $a/R = 10^5$ and for the Sun–Earth combination, we have $a/R_S = 200$, which value for the Sun–Jupiter combination becomes 1000. The corresponding volume ratio is $(a/R_S)^3$. This becomes 10^{15} for the atom and 10^9 for the Sun–Jupiter combination. We can conclude, therefore, that the solar system is more condensed than the atomic world by a factor of 10^6.

Comparing revolutions, the hydrogen atom in ground state performs 10^{16} rev/s, corresponding to 3×10^{23} rev/year, while Earth makes 1 rev/year and the corresponding value for Jupiter is 0.0843 rev/year.

So the 10^6 times denser solar system moves much slower, in fact by a factor of 3.57×10^{24}, considering the number of revolutions performed by the two systems during the same time. Because of the popularity of supercolliders (i.e., accelerators), their corresponding 6×10^{10} rev/year might also be mentioned. Considering the age of the solar system (about 10^9 years), the corresponding running time for an accelerator is 6 days. Conclusions are left to the reader.

2.10. This example uses the idea of the potential function to find the gravitational potential and the force acting on a point mass m by two equal masses (M) separated by a distance $2D$ as shown in Figure 2.10. The gravitational force field of the two equal masses is best determined by using the potential function evaluated at the position of the mass m:

$$V_g = GmM\left(\frac{1}{r_1} + \frac{1}{r_2}\right)$$

The force acting on m is the gradient of the potential,

$$\mathbf{grad}\ V_g = \hat{r}\,\frac{\partial V_g}{\partial r} + \hat{\theta}\,\frac{1}{r}\,\frac{\partial V_g}{\partial \theta}$$

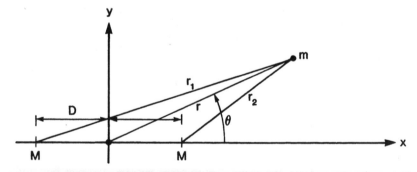

FIGURE 2.10 Gravitational force field of two equal masses.

where $\hat{\theta}$ is the unit vector normal to $\mathbf{r}$ in the xy plane. Once r_1 and r_2 are expressed as functions of r and θ, we have

$$V_g = GmM \left(\frac{1}{\sqrt{r^2 + D^2 - 2rd \cos \theta}} + \frac{1}{\sqrt{r^2 + D^2 + 2rD \cos \theta}} \right)$$

If $D/r \ll 1$, the terms under the square roots can be approximated, and we have

$$V_g = \frac{GmM}{r} \left(2 - \frac{D^2}{r^2} \sin^2 \theta \right)$$

The corresponding force becomes

$$\mathbf{F}_g = \mathbf{grad}\, V_g = -2GmM \left[\hat{r} \left(\frac{1}{r^2} - \frac{3}{2} \frac{\sin^2 \theta}{r^4} \right) + \hat{\theta}\, \frac{\sin \theta \cos \theta}{r^4} \right]$$

PROBLEMS

2.1. Find the mean motion, altitude, and circular velocity of a geosynchronous satellite using the precise period of the sidereal day: 23 h, 56 min, 4.1 s.

2.2. Derive linear formulas showing the analytical dependence of the deviations in ω, h, and v_{cs} on a small change in the period: $\Delta \omega = f_1(\Delta T)$, $\Delta h = f_2(\Delta T)$, and $\Delta v_{cs} = f_3(\Delta T)$.

2.3. Compare the actual values of the semi-major axes of the planets as given in the table of constants of the planets in Appendix 2 with the values computed from Bode's law (actually announced by J. D. Titius in 1766), according to which

$$a_n = (4 + 3 \times 2^n) \times 0.1$$

Here, a_n is the distance of the nth planet in astronomical units from the Sun. Note that $n = -\infty, 0, 1, 2, 3, \ldots$, and the value $n = 3$ corresponds to the orbits of minor planets, such as Ceres, discovered in 1801 with the help of the *Titius–Bode law*. *Note:* The astronomical unit is the distance from Earth to the Sun.

CHAPTER 3

THE GENERAL PROBLEM
OF TWO BODIES

In Chapter 2, we looked at the special case of circular orbits and derived some important practical applications. These are of interest because many orbits, both in celestial mechanics and in the motion of artificial satellites around Earth, are almost circular. In doing this, however, we made some assumptions that require more rigorous mathematical proof. It will turn out that circular orbits are indeed permitted and that these are a special case of the general class of orbits characteristic of bodies moving according to the law of gravitational attraction and Newton's law of motion. In Chapter 1, we wrote down the law of gravitation [equation (1.2)] as derived by Newton:

$$\mathbf{F} = -\frac{GMm}{r^2}\,\hat{r} \qquad (3.1)$$

which is a vector equation that defines the force $\mathbf{F}$ acting between the two masses m and M separated by the distance r. The equation says that the gravitational force is proportional to the product of the two masses, that it is inversely proportional to the square of the distance between them, and that it points along the line joining the two masses, which is indicated by the unit vector $\hat{r}$. The proportionality constant G is called the gravitational constant, and is one of the "universal" constants of nature. (The numerical value of G is 6.672×10^{-11} m³/kg s².) The negative sign that appears in equation (3.1) is a mathematical convention we have

33

added to indicate that the gravitational force between the two masses is always attractive.

Newton's laws of motion were also stated in Chapter 1. The most important of these is the "second law," which can be expressed as

$$F = \frac{d(m\mathbf{v})}{dt} = m\frac{d\mathbf{v}}{dt} \tag{3.2}$$

and is identical to equation (1.1). This equation says that the change in the momentum, $m\mathbf{v}$, of a body is proportional to the force imposed, $\mathbf{F}$, multiplied by the time interval dt during which the force acts. The velocity of the body, $\mathbf{v}$, is also a vector, and it can be written as follows in terms of the position vector $\mathbf{r}$ of the body:

$$\mathbf{v} = \frac{d\mathbf{r}}{dt} \tag{3.3}$$

Combining equations (3.2) and (3.3) leads to

$$\mathbf{F} = m\frac{d^2\mathbf{r}}{dt^2} \tag{3.4}$$

which is the familiar statement of Newton's second law of motion. The first law of motion, stated in Chapter 1, is a consequence of equation (3.4) for the case $\mathbf{F} = 0$. The third law of motion is a statement of the conservation of momentum to which we will return shortly.

By combining equations (3.1) and (3.4), we obtain the equation that governs the general motion of two bodies interacting through a gravitational force:

$$m\frac{d^2\mathbf{r}}{dt^2} = -\frac{GmM}{r^2}\hat{r} \tag{3.5}$$

The coordinate system in which equation (3.5) is written is somewhat obscure. Actually, the three variables that define the motion of the two bodies are the magnitude of the vector $\mathbf{r}$ and the two angles that define the direction of the unit vector $\hat{r}$. These variables will be dealt with in more detail when the actual solutions of equation (3.5) are developed. In addition to being a vector equation, equation (3.5) is also nonlinear. Nonlinear differential equations are often not solvable by analytic methods. By a great stroke of good fortune, it turns out that equation (3.5) is "integrable," that is, a way exists to obtain exact integrals of this equation that permit exact solutions. Were it not for that, Newton could not have established the theory of orbits in the seventeenth century. We would have had to wait until the advent of high-speed computers to obtain numerical solutions.

In order to examine the nature of equation (3.5) in greater detail, it is worthwhile to look at the equation in terms of the variables shown in Figure 3.1. The Cartesian coordinate system shown in Figure 3.1 is a reference system that will be employed in subsequent chapters to write the equations of the orbits both in Cartesian and in polar coordinates. Figure 3.1 is simply an extension of Figure 2.1 in that is uses an arbitrary origin for the coordinate system at the point O rather than at the center of mass of the system. While nothing other than equation (3.5) will actually result from the exercise of using an arbitrarily located origin in the case of two bodies, it is important to develop the formalism because it is indeed necessary to go through such an analysis when the motion of three or more bodies interacting through gravity is considered. (We will return to this point in Chapter 12.)

Figure 3.1 shows two masses, m_1 and m_2, located at $\mathbf{r}_1$ and $\mathbf{r}_2$ in the coordinate system with the origin at the point O. Equation (3.5) can now be rewritten to define the force exerted on mass m_1 by the mass m_2:

$$\mathbf{F}_{12} = m_1 \frac{d^2 \mathbf{r}_1}{dt^2} = -\frac{Gm_1 m_2}{|\mathbf{r}_1 - \mathbf{r}_2|^3}(\mathbf{r}_1 - \mathbf{r}_2) \qquad (3.6)$$

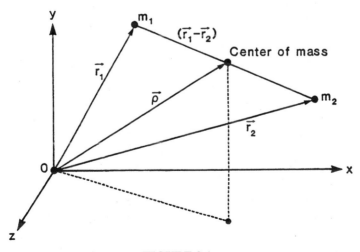

FIGURE 3.1

The force on the mass m_2 due to m_1 will be equal and opposite to $\mathbf{F}_{12}$:

$$\mathbf{F}_{21} = -\mathbf{F}_{12} = m_2 \frac{d^2\mathbf{r}_2}{dt^2} = \frac{Gm_1m_2}{|\mathbf{r}_1 - \mathbf{r}_2|^3}\left(\mathbf{r}_1 - \mathbf{r}_2\right) \tag{3.7}$$

Adding equations (3.6) and (3.7) yields

$$m_1 \frac{d^2\mathbf{r}_1}{dt^2} + m_2 \frac{d^2\mathbf{r}_2}{dt^2} = 0 \tag{3.8}$$

Equation (3.8) can be integrated to give

$$m_1 \frac{d\mathbf{r}_1}{dt} + m_2 \frac{d\mathbf{r}_2}{dt} = \mathbf{P} \tag{3.9}$$

where $\mathbf{P}$ is a constant having the dimensions of a momentum vector. This can be related to the velocity of the center of mass of the system of two masses, m_1 and m_2, as follows: By the definition of the center of mass, we have

$$(m_1 + m_2)\,\boldsymbol{\rho} = m_1\mathbf{r}_1 + m_2\mathbf{r}_2 \tag{3.10}$$

where the vector $\boldsymbol{\rho}$ is the position of the center of mass as defined in Figure 3.1. We can rewrite equation (3.9) as follows:

$$\mathbf{P} = \frac{d}{dt}(m_1\mathbf{r}_1 + m_2\mathbf{r}_2) = (m_1 + m_2)\frac{d\boldsymbol{\rho}}{dt} \tag{3.11}$$

where $d\boldsymbol{\rho}/dt$ is the velocity of the center of mass, $\mathbf{v}_{CM}$:

$$\mathbf{v}_{CM} = \frac{d\boldsymbol{\rho}}{dt} = \frac{\mathbf{P}}{m_1 + m_2} \tag{3.12}$$

Note that this velocity is constant. This means that there are no external forces acting on the center of mass. Thus, we can understand the two-body problem by looking only at the *relative motion* of the two masses with respect to the center of mass.

Having established the point that we are dealing only with the relative motion of the two masses, we can now rewrite the equation of motion in terms of the vector $\mathbf{r}_1 - \mathbf{r}_2$, which is the variable that describes the motion. Subtracting equation (3.7) from equation (3.6) yields

$$\frac{d^2\mathbf{r}_1}{dt^2} - \frac{d^2\mathbf{r}_2}{dt^2} = \frac{-G(m_1 + m_2)}{|\mathbf{r}_1 - \mathbf{r}_2|^3}(\mathbf{r}_1 - \mathbf{r}_2) \qquad (3.13)$$

or

$$\frac{d^2}{dt^2}(\mathbf{r}_1 - \mathbf{r}_2) = -\frac{G(m_1 + m_2)}{|\mathbf{r}_1 - \mathbf{r}_2|^3}(\mathbf{r}_1 - \mathbf{r}_2) \qquad (3.14)$$

Equation (3.14) is the fundamental equation that defines the motion of the masses m_1 and m_2 with respect to each other under the influence of gravitational forces. It can be seen that equation (3.14) has the same form as equation (3.5) by referring to Figure 2.1 in the previous chapter. In this case, the vector $\boldsymbol{\rho}$ is equal to zero. It is also clear from Figure 2.1 that the magnitude of the vector $\mathbf{r}_1 - \mathbf{r}_2$ is $r_1 + r_2$, or

$$|(\mathbf{r}_1 - \mathbf{r}_2)| = r_1 + r_2 \qquad (3.15)$$

The position of mass m_1 with respect to the coordinate system of Figure 2.1 is $\mathbf{r}_1$. Multiplying equation (3.14) by m_1 and then using equation (3.15), we obtain

$$m_1\frac{d^2\mathbf{r}_1}{dt^2} = -\frac{G(m_1 + m_2)m_1}{(r_1 + r_2)^3}(\mathbf{r}_1 - \mathbf{r}_2) \qquad (3.16)$$

Note also that the vector $\mathbf{r}_1 - \mathbf{r}_2$ can be rewritten as follows:

$$(\mathbf{r}_1 - \mathbf{r}_2) = (r_1 + r_2)\hat{r} \qquad (3.17)$$

where $\hat{r}$ is the unit vector that defines the direction in which the vector $\mathbf{r}_1 - \mathbf{r}_2$ points. Thus, we can write

$$m_1\frac{d^2\mathbf{r}_1}{dt^2} = -\frac{Gm_1M_R}{r_1^2}\hat{r}_1 \qquad (3.18)$$

From the definition of the center of mass in Figure 2.2, we have

$$m_1r_1 = m_2r_2 \text{ or } r_2 = \frac{m_1}{m_2}r_1 \qquad (3.19)$$

(*Note:* We have placed m_E and m_S in Figure 2.1 with m_1 and m_2 to write this equation.) Substituting equation (3.19) in (3.18) yields

$$m_1 \frac{d^2\mathbf{r}_1}{dt^2} = \frac{-Gm_1(m_1 + m_2)}{[r_1 + (m_1/m_2)r_1]^2} \hat{r} = -\frac{Gm_1m_2^2}{m_1 + m_2} \frac{\hat{r}}{r_1^2} \qquad (3.20)$$

This equation has precisely the same formation as equation (3.5) if the quantity M in that equation is defined as M_R:

$$M_R = \frac{m_2^2}{m_1 + m_2} \qquad (3.21)$$

Equation (3.20) states that the mass m_1 moves around the center of mass of the system as if the mass M_R defined in equation (3.21) were located at the center of mass. The quantity M_R is called the "reduced mass" of the system if the motion of m_1 is considered. It is obvious that a similar argument can be made to calculate the motion of mass m_2, in which case the reduced mass is given as

$$M_R = \frac{m_1^2}{m_1 + m_2} \qquad (3.22)$$

Thus, we have the following result. The equation of motion for the mass m_1 is

$$m_1 \frac{d^2\mathbf{r}_1}{dt^2} = -\frac{Gm_1M_R(\text{mass } m_1)}{r_1^2} \hat{r} \qquad (3.23)$$

and for mass m_2

$$m_2 \frac{d^2\mathbf{r}_2}{dt^2} = -\frac{Gm_2M_R(\text{mass } m_2)}{r_2^2} \hat{r} \qquad (3.24)$$

Note that the reduced mass M_R is always smaller than m_2 or m_1, depending on which mass is being considered. If one of the masses is very much larger than the other, an important simplification results. If we assume that $m_2 \gg m_1$, then the reduced mass of the system is

$$M_R \approx m_2 \qquad (3.25)$$

The equation of motion for m_1 then becomes

$$m_1 \frac{d^2\mathbf{r}_1}{dt^2} = -\frac{Gm_1m_2}{r_1^2}\hat{r} \qquad (3.26)$$

which results when equation (3.25) is substituted into (3.20). Equation (3.26) is precisely the same as equation (3.5) with $m = m_1$ and $M = m_2$. The approximation that $m_2 \gg m_1$ assumes essentially that the center of mass is located at mass m_2. This is a very useful approximation because it applies to many practical cases. Earth moving around the Sun, the Moon moving around Earth, and an artificial satellite moving around the Sun all fit the approximation that $m_2 \gg m_1$.

There are two other important results that follow from the equations of motion for m_1 or m_2: (3.23) and (3.24). We will assume that the mass m_1 is the mass we consider to be moving so that the equation of motion is

$$m_1 \frac{d^2\mathbf{r}_1}{dt^2} = -\frac{Gm_1M_R}{r_1^2}\hat{r}$$

or

$$m_1 \frac{d^2\mathbf{r}_1}{dt^2} = -\frac{Gm_1M_R}{r_1^3}\mathbf{r}_1 \qquad (3.27)$$

This is a vector equation that can be integrated once by the following procedure: Multiply both sides by the velocity vector $d\mathbf{r}_1/dt$:

$$m_1 \frac{d\mathbf{r}_1}{dt}\cdot\frac{d^2\mathbf{r}_1}{dt^2} = -\frac{Gm_1M_R}{r_1^3}\mathbf{r}_1\cdot\frac{d\mathbf{r}_1}{dt} \qquad (3.28)$$

This equation is a scalar equation that can be simplified by the following procedures:

$$\frac{d}{dt}(\mathbf{r}_1\cdot\mathbf{r}_1) = \mathbf{r}_1\cdot\frac{d\mathbf{r}_1}{dt} + \frac{d\mathbf{r}_1}{dt}\cdot\mathbf{r}_1 = 2\mathbf{r}_1\cdot\frac{d\mathbf{r}_1}{dt} \qquad (3.29)$$

and likewise,

$$\frac{d}{dt}\left(\frac{d\mathbf{r}_1}{dt}\cdot\frac{d\mathbf{r}_1}{dt}\right) = 2\frac{d\mathbf{r}_1}{dt}\cdot\frac{d^2\mathbf{r}_1}{dt^2} \qquad (3.30)$$

Using equation (3.30), the left side of equation (3.28) becomes

$$m_1 \frac{d\mathbf{r}_1}{dt} \cdot \frac{d^2\mathbf{r}_1}{dt^2} = \frac{1}{2} m_1 \frac{d}{dt}\left(\frac{d\mathbf{r}_1}{dt} \cdot \frac{d\mathbf{r}_1}{dt} \right) \tag{3.31}$$

To rewrite the right side of equation (3.27), we need to look at the variable r_1 and its derivatives. Specifically,

$$\frac{d}{dt}\left(\frac{1}{r_1} \right) = -\frac{1}{r_1^2} \frac{dr_1}{dt} \tag{3.32}$$

We have to relate this to the vector $\mathbf{r}_1$ by writing

$$r_1 = (\mathbf{r}_1 \cdot \mathbf{r}_1)^{1/2} \tag{3.33}$$

so that

$$\frac{dr_1}{dt} = \frac{d}{dt}(\mathbf{r}_1 \cdot \mathbf{r}_1)^{1/2} = \frac{1}{2} \frac{1}{(\mathbf{r}_1 \cdot \mathbf{r}_1)^{1/2}} 2\mathbf{r}_1 \cdot \frac{d\mathbf{r}_1}{dt} \tag{3.34}$$

and

$$\frac{dr_1}{dt} = \frac{1}{r_1} \mathbf{r}_1 \cdot \frac{d\mathbf{r}_1}{dt} \tag{3.35}$$

Therefore, equation (3.32) can be rewritten as

$$\frac{d}{dt}\left(\frac{1}{r_1} \right) = -\frac{1}{r_1^3} \mathbf{r}_1 \cdot \frac{d\mathbf{r}_1}{dt} \tag{3.36}$$

Substituting (3.36) into the right-hand side of (3.28), we obtain

$$m_1 \frac{d\mathbf{r}_1}{dt} \cdot \frac{d^2\mathbf{r}_1}{dt^2} = Gm_1 M_R \frac{d}{dt}\left(\frac{1}{r_1} \right) \tag{3.37}$$

The right-hand side of equation (3.37) can be rewritten using equation (3.31), which yields

$$\frac{d}{dt}\left[\frac{1}{2} m_1 \frac{d\mathbf{r}_1}{dt} \cdot \frac{d\mathbf{r}_1}{dt} \right] = Gm_1 M_R \frac{d}{dt}\left(\frac{1}{r_1} \right) \tag{3.38}$$

Equation (3.38) can be integrated to yield

$$\frac{1}{2}m_1\frac{d\mathbf{r}_1}{dt}\cdot\frac{d\mathbf{r}_1}{dt} = \frac{Gm_1M_R}{r_1} + K \tag{3.39}$$

where K is a constant of integration. The left side of equation (3.39) can be further simplified by writing

$$\mathbf{v}_1 = \frac{d\mathbf{r}_1}{dt} \tag{3.40}$$

which is the velocity vector of the mass m_1. Substituting (3.40) into (3.39) yields

$$\frac{Gm_1M_R}{r_1} = \tfrac{1}{2}m_1v_1^2 + K \tag{3.41}$$

Equation (3.41) is called the energy equation, and the interpretation of the constant of integrating now becomes clear. In a system of two bodies such as the one being considered, the total energy is conserved since there is no dissipative mechanism that would change the total energy. Thus, K is related to the total energy as follows:

$$E_T = -K \tag{3.42a}$$

so that

$$E_T = \tfrac{1}{2}m_1v_1^2 - \frac{Gm_1M_R}{r_1} \tag{3.42b}$$

The first term on the right side of equation (3.42b) is the kinetic energy of the mass m_1, and the second term is the potential energy, as in equation (2.21). This equation defines the potential energy between m_1 and M_R separated by a distance r_1. Note that the total energy E_T that results from the initial conditions of the motion can be either positive or negative. It will be shown subsequently that the type of orbit executed by the mass m_1 depends upon the total energy E_T as defined here.

There is a second constant of the motion that is also important. This one can also be derived from equation (3.27). We start by forming the vector product of both sides of that equation with the position vector $\mathbf{r}_1$ of the mass m_1:

$$\mathbf{r}_1 \times m_1 \frac{d^2\mathbf{r}_1}{dt^2} = -\frac{Gm_1 M_R}{r_1^3}\mathbf{r}_1 \times \mathbf{r}_1 \tag{3.43}$$

Because of the properties of the vector product, the right-hand side of equation (3.43) is zero. The left-hand side of the equation can be rewritten by using the "chain rule" for vector products:

$$\frac{d}{dt}\left(\mathbf{r}_1 \times \frac{d\mathbf{r}_1}{dt}\right) = \frac{d\mathbf{r}_1}{dt} \times \frac{d\mathbf{r}_1}{dt} + \mathbf{r}_1 \times \frac{d^2\mathbf{r}_1}{dt} \tag{3.44}$$

Again, the first term on the right side is zero so that, using (3.43), we obtain

$$m_1 \frac{d}{dt}\left(\mathbf{r}_1 \times \frac{d\mathbf{r}_1}{dt}\right) = 0 \tag{3.45}$$

Integrating this equation yields

$$m_1\mathbf{r}_1 \times \frac{d\mathbf{r}_1}{dt} = \mathbf{L} \tag{3.46}$$

where the constant of integration is the vector $\mathbf{L}$. The same relationship as (3.46) holds for $\mathbf{r}_2$ and m_2.

The first point about the result expressed in equation (3.46) is that the vector $\mathbf{L}$ has both a constant magnitude and a constant direction. The latter defines a plane (perpendicular to the direction of $\mathbf{L}$) in which the motion of the masses m_1 and m_2 takes place. Thus, the motion takes place in two rather than in three dimensions. Having established this point, it is useful to describe the motion of m_1 and the origin O in terms of polar coordinates. Note that the vector $d\mathbf{r}_1/dt$ is the velocity of the mass m_1 with respect to O:

$$\mathbf{v}_1 = \frac{d\mathbf{r}_1}{dt} \tag{3.47}$$

Figure 3.2 defines the polar coordinate system in terms of the variables r and θ. Equation (3.46) can now be rewritten as

$$m_1\mathbf{r}_1 \times \mathbf{v}_1 = \mathbf{L} \tag{3.48}$$

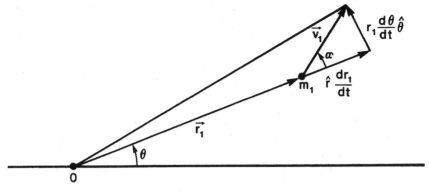

FIGURE 3.2

From Figure 3.2 and the definition of the vector product, we have, for the magnitude of the vector $\mathbf{r}_1 \times \mathbf{v}_1$,

$$|\mathbf{r}_1 \times \mathbf{v}_1| = r_1 v_1 \sin \alpha \qquad (3.49)$$

where α is the angle between $\mathbf{r}_1$ and $\mathbf{v}_1$. Furthermore, from the geometry of Figure 3.2, we have

$$v_1 \sin \alpha = r_1 \frac{d\theta}{dt} \qquad (3.50)$$

where $v_1 = |\mathbf{v}_1|$ and $r_1 = |\mathbf{r}_1|$. Using equations (3.40) and (3.49), we can rewrite equation (3.46) as

$$\mathbf{L} = m_1 r_1^2 \frac{d\theta}{dt} \hat{l} \qquad (3.51)$$

The quantity $m_1 r_1^2$ is the moment of inertia of a point of mass m_1 moving around the origin and $d\theta/dt$ is the angular velocity. Thus, $\mathbf{L}$ is the angular momentum vector and $\hat{l}$ is the unit vector that defines the direction in which it points. This result states that the angular momentum, along with the total energy, is also a constant of the motion for a general system of two masses interacting through the force of gravity. The conservation of angular momentum is the result of the fact that the gravitational force points in the direction of the line joining the two masses. Thus, the gravitational force cannot exert any torque that could change the angular momentum because the moment arm is always zero.

For further reading on the material in this chapter, the introductory chapters in the books of Danby (1962) and Roy (1978) are recommended. (See references listed in the Appendix.)

EXAMPLES

3.1. The solution of the problem of two bodies can be represented by a Taylor series. This solution is known in the literature as the f and g series.

We write the solution as

$$\mathbf{r}(t) = \mathbf{r}_0 f + \frac{d\mathbf{r}_0}{dt} g$$

where the functions f and g depend on the time and on the initial conditions and $\mathbf{r}_0$ is the value of the vector at $t = t_0$.

The Taylor series solution is

$$\mathbf{r}(t) = \mathbf{r}_0 + \frac{d\mathbf{r}_0}{dt}(t - t_0) + \frac{1}{2}\frac{d^2\mathbf{r}_0}{dt^2}(t - t_0)^2 + \frac{1}{6}\frac{d^3\mathbf{r}_0}{dt^3}(t - t_0)^3 + \ldots$$

The initial conditions are represented by the position and velocity vectors at $t = t_0$, which are denoted by $\mathbf{r}_0$ and $d\mathbf{r}_0/dt$. All other coefficients of the series can be expressed by these vectors, as shown by the following derivation. From the equation of motion, we have

$$\frac{d^2\mathbf{r}}{dt^2} = -\frac{G(m_1 + m_2)}{|\mathbf{r}|^3}\mathbf{r}$$

corresponding to equation (3.14) with $\mathbf{r} = (\mathbf{r}_1 - \mathbf{r}_2)$.

At $t = t_0$, the above equation becomes

$$\frac{d^2\mathbf{r}_0}{dt^2} = -\frac{G(m_1 + m_2)}{|\mathbf{r}_0|^3}\mathbf{r}_0$$

In this way, the third term of the series is obtained. The fourth term requires the computation of the third derivative of $\mathbf{r}$ from the equation of motion:

$$\frac{d^3\mathbf{r}}{dt^3} = -G(m_1 + m_2)\frac{|\mathbf{r}|^3(d\mathbf{r}/dt) - \mathbf{r}(d|\mathbf{r}|^3/dt)}{|\mathbf{r}|^6}$$

Here the derivative of $|\mathbf{r}|^3$ requires some attention, but its equivalent is not more difficult than the previously discussed evaluations of time derivatives. We have

$$\frac{d|\mathbf{r}|^3}{dt} = \frac{d[\mathbf{r}\cdot\mathbf{r}]^{3/2}}{dt} = \frac{3}{2}\left(\frac{1}{\mathbf{r}\cdot\mathbf{r}}\right)^{1/2}\frac{d(\mathbf{r}\cdot\mathbf{r})}{dt} = \frac{3}{2}|\mathbf{r}|2\mathbf{r}\frac{d\mathbf{r}}{dt}$$

The third derivative becomes

$$\frac{d^3\mathbf{r}}{dt^3} = -G(m_1 + m_2)\frac{|\mathbf{r}|^3(d\mathbf{r}/dt) - \mathbf{r}[3\mathbf{r}\cdot(d\mathbf{r}/dt)|\mathbf{r}|]}{|\mathbf{r}|^6}$$

or

$$\frac{d^3\mathbf{r}}{dt^3} = -G(m_1 + m_2)\left[\frac{1}{|\mathbf{r}|^3}\frac{d\mathbf{r}}{dt} - 3\left(\mathbf{r}\cdot\frac{d\mathbf{r}}{dt}\right)\frac{\mathbf{r}}{|\mathbf{r}|^5}\right]$$

At $t = t_0$ we have

$$\frac{d^3\mathbf{r}_0}{dt^3} = -G(m_1 + m_2)\left[\frac{1}{|\mathbf{r}_0|^3}\frac{d\mathbf{r}_0}{dt} - 3\left(\mathbf{r}_0\cdot\frac{d\mathbf{r}_0}{dt}\right)\frac{\mathbf{r}_0}{|\mathbf{r}_0|^5}\right]$$

The Taylor series expansion becomes

$$\mathbf{r}(t) = \mathbf{r}_0 + \frac{d\mathbf{r}_0}{dt}(t - t_0) - \frac{G(m_1 + m_2)}{2}\frac{\mathbf{r}_0}{|\mathbf{r}_0|^3}(t - t_0)^2$$

$$+ \left[\frac{G(m_1 + m_2)}{2}\left(\mathbf{r}_0\cdot\frac{d\mathbf{r}_0}{dt}\right)(t - t_0)\frac{\mathbf{r}_0}{|\mathbf{r}_0|^5} - \frac{G(m_1 + m_2)}{6}\frac{1}{|\mathbf{r}_0|^3}\frac{d\mathbf{r}_0}{dt}\right]$$

$$\times (t - t_0)^3 + \ldots$$

The members of the f series have the factor $\mathbf{r}_0$ and those of the g series are multiplied by $d\mathbf{r}_0/dt$:

$$f = 1 - \frac{G(m_1 + m_2)}{2}\frac{(t - t_0)^2}{r_0^3} + \frac{G(m_1 + m_2)}{2}\frac{[\mathbf{r}_0\cdot(d\mathbf{r}_0/dt)]}{r_0^5}(t - t_0)^3 \pm \ldots$$

$$g = (t - t_0) - \frac{G(m_1 + m_2)}{6r_0^3}(t - t_0)^3 \pm \ldots$$

where $r_0 = |\mathbf{r}_0|$ is the length of the initial position vector.

Note that the second factor of the third term of the f series can be written as

$$\mathbf{r}_0 \cdot \frac{d\mathbf{r}_0}{dt} = |\mathbf{r}_0| \left| \frac{d\mathbf{r}_0}{dt} \right| \cos \alpha$$

where

$$\frac{d\mathbf{r}_0}{dt} = \left| \frac{d\mathbf{r}_0}{dt} \right| \cos \alpha$$

is the initial value of the radial velocity component (see Figure 3.2). This notation can be misleading since $|\mathbf{r}_0| = r_0$ is the length of the initial radial position vector, but $|d\mathbf{r}_0/dt| \neq d\mathbf{r}_0/dt$ as mentioned before.

With this notation, the f series becomes

$$f = 1 - \frac{G(m_1 + m_2)}{2} \frac{(t - t_0)^2}{r_0^3} + \frac{G(m_1 + m_2)}{2} \frac{1}{r_0^4} \frac{d\mathbf{r}_0}{dt} (t - t_0)^3 \pm \ldots$$

For short times, these series are the basic tools of orbit determination. The convergence properties are discussed in Chapter 8 of Taff's book (see Appendix). The first 35 terms of the series are given in A. Deprit's "Fundamentals of Astrodynamics," Mathematical Note No. 556, Boeing Scientific Research Laboratories, Seattle, WA, 1968.

PROBLEMS

3.1. Find the approximate distances between the Sun and the centers of mass of the Sun–Jupiter, Sun–Saturn, and Sun–Earth systems. Find the approximate circular speed of the center of the Sun in these systems.

3.2. Find the approximate distances between Earth and the centers of mass of the Earth–Moon system, and find the approximate circular speed of the center of Earth in this system.

CHAPTER 4

ELLIPTIC ORBITS

In this chapter, we shall show that two bodies moving around their common center of mass execute elliptic orbits under certain conditions. As in Chapter 3, we will call the masses m_1 and m_2, and we will start with equation (3.26) of that chapter:

$$m_1 \frac{d^2\mathbf{r}}{dt^2} = \frac{-Gm_1m_2}{r_1^2}\hat{r} \qquad (4.1)$$

where the quantity r_1 is defined in Figure 3.1 and the unit vector $\hat{r}$ points in the direction defined by the line joining m_1 and m_2.

To find a solution of the vector equation (4.1), we will refer to the coordinate system defined in Figure 4.1. The mass m_2 is considered to be at the origin of the coordinate system. This assumption simplifies the derivation, and we can either say that m_2 represents the "reduced mass," as defined in equation (3.21), of m_1 and some other mass m_2 or that $m_1 \ll m_2$. In either case, the orbit to be determined will be executed around the origin of the coordinate system shown in Figure 4.1. To simplify things, we will drop the subscripts and call the test mass m and the mass at the origin, M. The equation to be solved is then

$$m \frac{d^2\mathbf{r}}{dt^2} = \frac{-GmM}{r^2}\hat{r} \qquad (4.2)$$

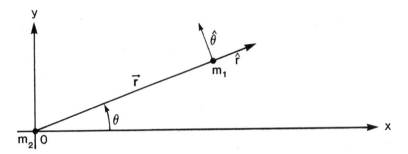

FIGURE 4.1

Equation (4.2) defines the behavior of the vector **r** as a function of time. In order first to determine the shape of the orbit, we want to eliminate the time variable and rewrite the equation that describes the orbit in terms of the variables r and θ described in the coordinate system of Figure 4.1. In order to calculate the second derivative on the left-hand side of equation (4.2), we start with the identity

$$\mathbf{r} = r\hat{r} \qquad (4.3)$$

where r is the magnitude of the vector **r**. The first derivative with respect to time is

$$\frac{d\mathbf{r}}{dt} = r\frac{d\hat{r}}{dt} + \hat{r}\frac{dr}{dt} \qquad (4.4)$$

The second derivative is

$$\frac{d^2\mathbf{r}}{dt^2} = r\frac{d^2\hat{r}}{dt^2} + \frac{dr}{dt}\frac{d\hat{r}}{dt} + \frac{d\hat{r}}{dt}\frac{dr}{dt} + \hat{r}\frac{d^2r}{dt^2} \qquad (4.5)$$

We must now look at the behavior of the unit vectors $\hat{r}$ and $\hat{\theta}$ as functions of time. The magnitude of the unit vectors clearly remains constant so that we need to be concerned only about changes in direction of the unit vectors. Since both unit vectors are attached to the vector **r**, as shown in Figure 4.1, they will rotate around the origin of the coordinate system with the angular velocity $d\theta/dt$ of the vector **r**. From Figure 4.1, it can be seen that the unit vector $\hat{r}$ changes in the $\hat{\theta}$ direction with a rate equal to the angular velocity of **r**:

$$\frac{d\hat{r}}{dt} = \hat{\theta}\frac{d\theta}{dt} \tag{4.6}$$

The unit vector $\hat{\theta}$ will also change but will rotate in a direction so that it always points to the origin of the coordinate system:

$$\frac{d\hat{\theta}}{dt} = -\hat{r}\frac{d\theta}{dt} \tag{4.7}$$

The negative sign in equation (4.7) reflects the fact that the change in the unit vector $\hat{\theta}$ always points toward the origin O. The second derivative of the unit vector $\hat{r}$ from equation (4.6) is

$$\frac{d^2\hat{r}}{dt^2} = \frac{d\hat{\theta}}{dt}\frac{d\theta}{dt} + \hat{\theta}\frac{d^2\theta}{dt^2} = -\hat{r}\left(\frac{d\theta}{dt}\right)^2 + \hat{\theta}\frac{d^2\theta}{dt^2} \tag{4.8}$$

Using equations (4.8) and (4.6) to rewrite equation (4.5), we obtain

$$\frac{d^2\mathbf{r}}{dt^2} = -\hat{r}r\left(\frac{d\theta}{dt}\right)^2 + \hat{\theta}r\frac{d^2\theta}{dt^2} + 2\hat{\theta}\frac{dr}{dt}\frac{d\theta}{dt} + \hat{r}\frac{d^2r}{dt^2}$$

$$= \hat{r}\left[\frac{d^2r}{dt^2} - r\left(\frac{d\theta}{dt}\right)^2\right] + \hat{\theta}\left(r\frac{d^2\theta}{dt^2} + 2\frac{dr}{dt}\frac{d\theta}{dt}\right) \tag{4.9}$$

The entire equation of motion (4.2) can now be written using equation (4.9):

$$\hat{r}\left[m\frac{d^2r}{dt^2} - mr\left(\frac{d\theta}{dt}\right)^2 + \frac{GmM}{r^2}\right] + \hat{\theta}\left(mr\frac{d^2\theta}{dt^2} + 2m\frac{dr}{dt}\frac{d\theta}{dt}\right) = 0 \quad (4.10)$$

This equation still contains time as a variable, and it is, of course, still a vector equation since equation (4.2) from which this is derived is also a vector relationship.

Equation (4.10) can now be simplified by using the fact that the angular momentum vector $\mathbf{L}$ defined in equation (3.46) is a constant both in magnitude and in direction. The magnitude of the vector $\mathbf{L}$ is given in equation (3.51) as

$$L = mr^2\frac{d\theta}{dt} \tag{4.11}$$

If we now take the time derivative of L, we obtain

$$\frac{dL}{dt} = 2mr\frac{dr}{dt}\frac{d\theta}{dt} + mr^2\frac{d^2\theta}{dt^2} = 0 \qquad (4.12)$$

The right side of equation (4.12) is equal to the second term in equation (4.10) so that the $\hat{\theta}$ component vanishes. Therefore, the $\hat{r}$ component must also vanish, which yields

$$m\frac{d^2r}{dt^2} - mr\left(\frac{d\theta}{dt}\right)^2 + \frac{GmM}{r^2} = 0 \qquad (4.13)$$

Using the angular momentum equation (4.11) again, this can be rewritten as

$$m\frac{d^2r}{dt^2} - \frac{L^2}{mr^3} + \frac{GmM}{r^2} = 0 \qquad (4.14)$$

Equation (4.14) defines how the magnitude of the vector $\mathbf{r}$ behaves as a function of time. We have already said that, initially, we will be interested not in the dynamics but rather in the shape, or better still, in the geometry of the orbit of mass m in the gravitational field of the mass M located at the point O. To do this, we must look at the quantity r not as a function of time but as a function of the angle θ. Therefore,

$$\frac{dr}{dt} = \frac{dr}{d\theta}\frac{d\theta}{dt} = \frac{L}{mr^2}\frac{dr}{d\theta} \qquad (4.15)$$

where we have once again used the angular momentum relationship (4.11). The second derivative is

$$\frac{d^2r}{dt^2} = -\frac{2L}{mr^3}\frac{dr}{dt}\frac{dr}{d\theta} + \frac{L}{mr^2}\frac{d^2r}{d\theta^2}\frac{d\theta}{dt} \qquad (4.16)$$

Using equations (4.11) and (4.15) to evaluate $d\theta/dt$ and dr/dt in terms of r and θ, we obtain

$$\frac{d^2r}{dt^2} = -\frac{2L^2}{m^2r^5}\left(\frac{dr}{d\theta}\right)^2 + \frac{L^2}{m^2r^4}\frac{d^2r}{d\theta^2} \qquad (4.17)$$

Using the expression in equation (4.17) for d^2r/dt^2 in equation (4.13) and using the angular momentum equation (4.11) to evaluate $d\theta/dt$, we obtain an equation that expresses r as a function of θ:

$$\frac{L^2}{mr^4}\frac{d^2r}{d\theta^2} - \frac{2L^2}{mr^5}\left(\frac{dr}{d\theta}\right)^2 - \frac{L^2}{mr^3} + \frac{GmM}{r^2} = 0 \qquad (4.18)$$

This equation can be simplified by collecting the coefficients so that the ultimate result looks like this:

$$\frac{1}{r^2}\frac{d^2r}{d\theta^2} - \frac{2}{r^3}\left(\frac{dr}{d\theta}\right)^2 - \frac{1}{r} + \frac{Gm^2M}{L^2} = 0 \qquad (4.19)$$

This is a nonlinear differential equation. As we have said in the previous chapter, nonlinear equations of this kind do not have general solutions that can be expressed as simple analytic functions. Fortunately, there is a simple transformation of variables that permits an exact analytic solution of equation (4.19):

$$r = \frac{1}{u} \qquad (4.20)$$

The derivatives of r with respect to θ can now be calculated it terms of u and θ using equation (4.20):

$$\frac{dr}{d\theta} = -\frac{1}{u^2}\frac{du}{d\theta} \qquad (4.21)$$

and the second derivative is

$$\frac{d^2r}{d\theta^2} = \frac{2}{u^3}\left(\frac{d}{d\theta^2}\right) - \frac{1}{u^2}\frac{d^2u}{d\theta^2} \qquad (4.22)$$

Using equations (4.22), (4.21), and (4.20) in (4.19), we have

$$u^2\left[\frac{2}{u^3}\left(\frac{du}{d\theta}\right)^2 - \frac{1}{u^2}\left(\frac{d^2u}{d\theta^2}\right)\right] - 2u^3\left[\frac{1}{u^4}\left(\frac{du}{d\theta}\right)^2\right] - u + \left(\frac{Gm^2M}{L^2}\right) = 0 \quad (4.23)$$

By further manipulation, equation (4.23) can be rewritten as follows:

$$\frac{d^2u}{d\theta^2} + u = \frac{Gm^2M}{L^2} \qquad (4.24)$$

Equation (4.24) has the same form as the equation describing a simple harmonic oscillator. The solutions of equation (4.24) can therefore be expressed as simple trigonometric functions. Specifically, the solution of (4.24) is

$$u = A \cos(\theta - \theta_0) + \frac{Gm^2M}{L^2} \tag{4.25}$$

which can easily be verified by differentiating equation (4.25) twice.

In terms of the actual variables that describe the geometry of the orbit, r and θ, we have

$$\frac{1}{r} = A \cos(\theta - \theta_0) + \frac{Gm^2M}{L^2} \tag{4.26}$$

where A and θ_0 are the constants of integration that have been carried over from the solution, equation (4.25). We will show shortly that these constants have specific physical meanings. Equation (4.26) can be rewritten as

$$r = \frac{1}{A \cos(\theta - \theta_0) + Gm^2M/L^2} \tag{4.27}$$

and, in order to see clearly physical interpretation of the constants appearing in (4.27), we rewrite the equation as

$$r = \frac{L^2/Gm^2M}{(AL^2/Gm^2M) \cos(\theta - \theta_0) + 1} \tag{4.28}$$

The usual form for this equation is

$$r = \frac{p}{e \cos(\theta - \theta_0) + 1} \tag{4.29}$$

where the quantity p is defined as

$$p = \frac{L^2}{Gm^2M} \tag{4.30}$$

and e is given as

$$e = \frac{AL^2}{Gm^2M} = Ap \tag{4.31}$$

The quantity p as defined by equation (4.30) is a constant determined by the two masses m and M, the gravitational constant G, and the value L of the angular momentum of the orbit. The quantity e depends on p but also on the constant of integration A, which defines the initial conditions of the motion of the mass m. The second constant of integration, θ_0, defines the orientation of the principal axis of the orbit with respect to an arbitrary external axis. (It will be demonstrated that the orbit actually has a principal axis.)

We shall now show that the orbit defined by the equation (4.29) defines an ellipse in the polar coordinates r and θ. The quantity e is called the eccentricity of the ellipse. In the case that the eccentricity is zero, equation (4.29) reduces to

$$r = p \tag{4.32}$$

which is the equation of a circle with the radius p. Now, we can use equation (4.30) to write

$$r = p = \frac{L^2}{Gm^2M} \tag{4.33}$$

The angular momentum L of a mass in a moving circular orbit with radius r is

$$L = mvr \tag{4.34}$$

where v is the velocity of the mass in the orbit. Combining (4.34) and (4.33), we can calculate the velocity of the mass m moving in a circular orbit of radius r around the mass M as

$$v^2 = \frac{GM}{r}$$

or

$$v = \sqrt{\frac{GM}{r}} \tag{4.35}$$

This is precisely the same expression that was derived in Chapter 2 for circular orbits [see equation (2.5)]. The quantity p is called the *semilatus rectum* of the ellipse in the case when the eccentricity of the orbit is not zero.

It is now important to relate the physical quantities G, m, and M to the parameters that define the elliptic orbit. In order to do this, we will refer to Figure 4.2. In this drawing, we have chosen the major axis of the ellipse to lie along the x axis. This means that we have set the integration constant θ_0 equal to zero. (We shall return to this point shortly when we define the "true anomaly.") The mass M is located at one focus, F, of the ellipse and the test mass m executes the orbit. The origin of the coordinate system used to describe the motion is also at the focus F at which the mass M is located. The length l is the distance from the focus to the center of the ellipse, the length a is called the semimajor axis, and b is the semiminor axis.

From Figure 4.2 it is now possible to derive the following relationships using equation (4.29) with θ_0 equal to zero:

$$a + l = \frac{p}{1-e} \qquad a - l = \frac{p}{1+e} \tag{4.36}$$

Combining these equations yields

$$2a = \frac{p}{1-e} + \frac{p}{1+e} = \frac{2p}{1-e^2}$$

or $\hphantom{xxxxxxxxxxxxxxxxxxxxxxxxxxxxxxxxxxxxxxxxxxxxxxxxxx}$ (4.37)

$$a = \frac{p}{1-e^2}$$

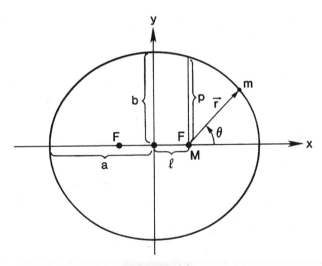

FIGURE 4.2

Also, by subtracting equation (4.36), we get

$$2l = \frac{p}{1-e} - \frac{p}{1+e} = \frac{2pe}{1-e^2} \qquad (4.38)$$

Combining this result with equation (4.37) yields

$$l = ae \qquad (4.39)$$

These are both useful relationships.

We shall now show that the equation of the ellipse in polar coordinates (4.29) is identical to the familiar equation for the ellipse in Cartesian coordinates when the appropriate transformation is made:

$$x = r \cos\theta \qquad y = r \sin\theta \qquad (4.39a)$$

Substituting (4.39a) into equation (4.29), with $\theta_0 = 0$, would result in the equation of an ellipse in a coordinate system with the origin at the right-hand focus of the ellipse:

$$p = r + er \cos\theta \qquad (4.40)$$

or, using (4.39),

$$p = \sqrt{x^2 + y^2} + ex \qquad (4.41)$$

Rewriting this equation yields

$$\sqrt{x^2 + y^2} = p - ex$$

and squaring both sides gives

$$x^2 + y^2 = p^2 - 2pex + e^2x^2 \qquad (4.42)$$

or

$$x^2(1 - e^2) + y^2 + 2pex = p^2 \qquad (4.43)$$

Completing the square on the left side of this equation and then doing some algebra yield

$$\left(x + \frac{pe}{1-e^2}\right)^2 + \frac{y^2}{1-e^2} = \frac{p^2}{(1-e^2)^2} \qquad (4.44)$$

We must now move the origin of the coordinate system to the center of the ellipse, which is accomplished by the transformation

$$x' = x + l = x + ae = x + \frac{pe}{1 - e^2} \qquad (4.45)$$

Substituting (4.45) into (4.44) yields

$$x'^2 + \frac{y^2}{1 - e^2} = \frac{p^2}{(1 - e^2)^2} \qquad (4.46)$$

and manipulating this equation to make it identical to the form of the equation of the ellipse in Cartesian coordinates yields

$$\frac{x'^2}{p^2/(1 - e^2)^2} + \frac{y^2}{p^2/(1 - e^2)} = 1 \qquad (4.47)$$

which is the same as the standard form of the Cartesian equation:

$$\frac{x'^2}{a^2} + \frac{y^2}{b^2} = 1 \qquad (4.48)$$

By comparing equations (4.47) and (4.48), we see that

$$a^2 = \frac{p^2}{(1 - e^2)^2} \quad \text{or} \quad a = \frac{p}{1 - e^2} \qquad (4.49)$$

This is identical to the relationship (4.37) that was derived from the geometric properties of the ellipse. Also, by comparing (4.48) and (4.47), we have

$$b^2 = \frac{p^2}{1 - e^2} \quad \text{or} \quad b = \frac{p}{\sqrt{1 - e^2}} \qquad (4.50)$$

The relationship between a and b is therefore

$$a = \frac{b}{\sqrt{1 - e^2}} \qquad (4.51)$$

We have now made the connection between the geometric properties of the ellipse and the physical quantities that define the orbit.

By proving that the orbit of a mass m moving around a fixed mass M is an ellipse, we have proved Kepler's first law of planetary motion (see Chapter 1). We have also shown that the other two laws are valid for circular orbits (see Chapter 2). We must now show that Kepler's laws hold for the general elliptic orbits that we have developed in this chapter. In order to do that, we must look not only at the geometric properties of the ellipse but also at the dynamics of the object moving in the elliptic orbit. This means that we have to reintroduce the time variable into the equations.

Kepler's second law, the "law of areas," follows from Figure 4.3. The area ΔA of the segment identified in Figure 4.3 is defined as

$$\Delta A = \tfrac{1}{2}(r^2 \, \Delta\theta)$$

or in the notation of differential calculus, this becomes

$$dA = \tfrac{1}{2}(r^2 \, d\theta) \tag{4.52}$$

Now dividing each side by dt, we have

$$\frac{dA}{dt} = \frac{1}{2}r^2\frac{d\theta}{dt} \tag{4.53}$$

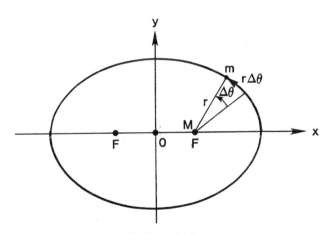

FIGURE 4.3

The term $r^2\, d\theta/dt$ on the right side, when multiplied by m, is the angular momentum that we have defined as the constant L:

$$\frac{dA}{dt} = \frac{1}{2}\frac{L}{m} = \text{const} \tag{4.54}$$

From (4.54) it can be seen that dA is always proportional to dt so that equal areas are swept out in equal times.

The third law of Kepler follows from a comparison of the geometric properties of the ellipse and the dynamics of the motion of the orbiting object. The area of an ellipse can be expressed as follows in terms of the semimajor and semiminor axes:

$$A = \pi ab \tag{4.55}$$

Integrating equation (4.53) results in

$$\int_0^A dA = A = \pi ab = \frac{1}{2}\frac{L}{m}\int_0^T dt = \frac{1}{2}\frac{L}{m}T$$

where T is the time that it takes the particle to go around the ellipse:

$$2\pi ab = \frac{L}{m}T \tag{4.56}$$

From equation (4.51), we have

$$b = a\sqrt{1 - e^2}$$

so that

$$2\pi a^2\sqrt{1 - e^2} = \frac{L}{M}T \tag{4.57}$$

We now have to eliminate $\sqrt{1 - e^2}$ because the relationship must be true for *any* ellipse. Squaring both sides of (4.57) gives

$$4\pi^2 a^4(1 - e^2) = \frac{L^2}{M^2}T^2 \tag{4.58}$$

Now,

$$1 - e^2 = \frac{p}{a} \tag{4.59}$$

and the quantity p is defined as

$$p = \frac{L^2}{Gm^2M} \tag{4.60}$$

Therefore,

$$1 - e^2 = \frac{L^2}{aGm^2M} \tag{4.61}$$

and using (4.61) in equation (4.58) yields

$$4\pi^2 a^4 \frac{L^2}{aGm^2M} = \frac{L^2}{m^2} T^2$$

or, by rearranging,

$$a^3 = \frac{GM}{4\pi^2} T^2 \tag{4.62}$$

This is Kepler's third law. Note that this equation has exactly the same form as equation (2.14), which is the third law derived for circular orbits. The only difference is that the radius R of the circle is replaced by the semimajor axis of the ellipse.

There is one final point that needs to be emphasized. In complex orbital situations, it sometimes happens that the coordinate system in which the ellipse has been defined is not the same as the coordinate system in which the entire problem is treated. In that case, θ_0 is not zero and there is a convention that is used to describe the situation, which is illustrated in Figure 4.4. The origin of the coordinate system is at the right-hand focus. The axis x is the axis coincident with the major axis of the ellipse. The angle θ_0 is the angle between x and x'. Since x' is the coordinate system in which the problem is to be solved, the variable θ is measured from the axis x'. The angle between the x axis and the vector $\mathbf{r}$ is called f, the *true anomaly*. It is obvious from Figure 4.4 that

$$f = \theta - \theta_0 \tag{4.63}$$

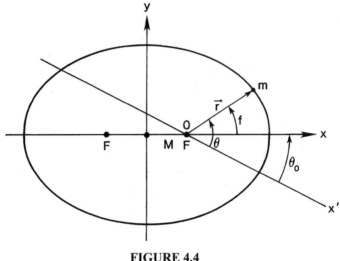

FIGURE 4.4

In such a situation, the true anomaly is the variable in the equation for the ellipse:

$$r = \frac{p}{1 + e \cos f} \qquad (4.64)$$

This will be useful when we study complex motions of satellites in orbit around Earth (see Chapter 11).

For further reading on the material in this chapter, the introductory chapters in the books of Danby (1962), McCuskey (1963) and Roy (1978) are recommended. (See references listed in the Appendix.)

EXAMPLES

4.1. Derive the radial (dr/dt) and normal to radial [$r(d\theta/dt)$] velocity components on elliptic orbits as functions of GM, a, e, and θ, with $\theta_0 = 0$. Using

$$r = \frac{p}{1 + e \cos \theta} \quad \text{and} \quad r^2 \frac{d\theta}{dt} = \frac{L}{M} = \sqrt{GMp}$$

where p is the semilatus rectum $p = a(1 - e^2)$, the radial velocity becomes

$$\frac{dr}{dt} = \sqrt{\frac{GM}{a}} \left(\frac{GM}{a(1 - e^2)} \right)^{1/2} (1 + e \cos \theta) \frac{dr}{dt} = \frac{dr}{d\theta} \frac{d\theta}{dt}$$

$$= e \left(\frac{GM}{a(1 - e^2)} \right)^{1/2} \sin \theta$$

Note that for a circular orbit $e = 0$ and the radial velocity becomes zero, as expected. The normal velocity component is

$$r \frac{d\theta}{dt} = \frac{\sqrt{GMp}}{r} \left(\frac{GM}{a(1 - e^2)} \right)^{1/2} (1 + e \cos \theta)$$

For a circular orbit, the normal velocity component becomes the circular velocity,

$$r \frac{d\theta}{dt} = v_c = \sqrt{\frac{GM}{a}}$$

The total speed v on elliptic orbits is defined as the magnitude of the vector velocity:

$$v = \mathbf{v} = \left[\left(\frac{dr}{dt} \right)^2 + r^2 \left(\frac{d\theta}{dt} \right)^2 \right]^{1/2} = \left[\frac{GM}{a(1 - e^2)} \right]^{1/2} (1 + e^2 + 2e \cos \theta)^{1/2}$$

which again becomes the circular velocity for $e = 0$. The same result can be obtained from the integral of energy,

$$v^2 = GM \left(\frac{2}{r} - \frac{1}{a} \right)$$

4.2. The period, apogee height, and perigee height of satellite *Explorer 7* were 1.684 h, 664 miles, and 346 miles. Find the semimajor axis, semiminor axis, and semilatus rectum in kilometers, the perigee and apogee velocities in kilometers per second, and the eccentricity of the orbit.

Twice the value of the semimajor axis is the sum of the perigee and apogee distances, measured from the center of Earth. The corresponding formula is

$$2a = 2R_E + h_p + h_a$$

where h_p = 556.834 km, h_a = 1068.606 km and R_E = 6378.14 km. The semimajor axis becomes a = 7190.86 km.

The eccentricity can be computed from the formula

$$e = \frac{h_a - h_p}{2a}$$

This equation can be obtained using the definitions of the perigee and apogee distances,

$$r_p = h_p + R_E = a(1 - e) \qquad r = h_a + R_E = a(1 + e)$$

and computing

$$h_a - h_p = a(1 + e) - a(1 - e) = 2ae$$

The eccentricity becomes e = 0.03558.

The semiminor axis and the semilatus rectum are given by

$$b = a\sqrt{1 - e^2} \quad \text{and} \quad p = a(1 - e^2)$$

The values are b = 7186.306 km and p = 7181.757 km. The perigee and apogee velocities can be calculated using the conservation of momentum:

$$L = mv_a a = mv_e b$$

so that when the velocities are evaluated, we have

$$v_p = \sqrt{\frac{GM}{a}}\sqrt{\frac{1+e}{1-e}} = 7.715 \text{ km/s} \qquad v_a = \sqrt{\frac{GM}{a}}\sqrt{\frac{1-e}{1+e}} = 7.185 \text{ km/s}$$

4.3. Show that for elliptic orbits the semimajor axis (a) is larger than the semiminor axis (b), which in turn is larger than the semilatus rectum (p), which finally is larger than the distance between the focus and the pericenter (r_p).

The series of inequalities stated above can be written as

$$a \geq b \geq p \geq r_p$$

Here $b = a\sqrt{1 - e^2}$, $p = a(1 - e^2)$, $r_p = a(1 - e)$, and $0 \leq e \leq 1$. The inequalities become

$$a \geq a\sqrt{1 - e^2} \geq a(1 - e^2) \geq a(1 - e)$$

The first inequality, after a is canceled, becomes

$$1 \geq 1 - e^2 \quad \text{or} \quad 0 \geq -e^2$$

This last inequality becomes an equality for $e = 0$. The second inequality becomes

$$1 \geq \sqrt{1 - e^2} \quad \text{or} \quad 1 \geq 1 - e^2 \quad \text{or} \quad 0 \geq -e^2$$

after cancellation by $a\sqrt{1 - e^2}$ and squaring. This again becomes an inequality when $e = 0$. The third inequality becomes

$$1 + e \geq 1 \quad \text{or} \quad e \geq 0$$

We conclude that the inequalities described are correct for elliptic orbits and become equalities for circular orbits. For straight-line or rectilinear orbits ($e = 1$), we have $b = p = r_p = 0$.

PROBLEMS

4.1. The perigee and apogee altitudes of an artificial Earth satellite are 200 and 500 km. Find the values of a, b, p, e, v_p, and v_a.

4.2. Show that the maximum value of dr/dt for an elliptic orbit occurs at the intersection of the latus rectum with the orbit and find this maximum value.

CHAPTER 5

ROCKETS

The rocket is an ancient device that is mentioned in the literature for the first time around A.D. 1220 in a Chinese chronicle. The Chinese used incendiary rockets in several battles (Pien-king, Kai-fung-fu, and others) to defend themselves against the Mongol invasions of China in the thirteenth century. It is likely that rockets were actually developed much earlier, when the Chinese accidentally discovered gunpowder while trying to repeat some Greek experiments on combustibles. They substituted potassium nitrate for ordinary salt and created the world's first explosive. No one knows exactly when this happened, but it is known that the ancient Greeks did experiment with combustible mixtures before the birth of Christ.

The first serious military rockets were developed by the British military engineer Sir William Congreve. What he did was to substitute metal casings for the wood and cardboard ones used by the Chinese and thus created a much more robust and effective military weapon. Congreve's rockets were used for the first time in 1806, when the British bombarded the French port city of Boulogne during the Napoleonic wars. Probably the most famous incident involving the use of Congreve rockets was the bombardment of Fort McHenry in Baltimore Harbor by a British fleet in 1814. This engagement was observed by Francis Scott Key, who described the scene as "the rocket's red glare" providing the light by which the star-spangled banner would be seen through the night of bombardment. His poem was later used as the text for the American national anthem.

The first person to make a quantitative calculation of the performance of rockets was a Russian mathematics professor, Konstantin Edouardovitch Tsiolkovsky. In 1903, he published the first derivation of what we now call the "rocket equation" in a Russian scientific journal, *Nautschnoje Obozrenije*. Unfortunately, no one in the West picked this up at the time. In 1919, Robert H. Goddard in a paper entitled, "A Method for Reaching High Altitudes," published by the Smithsonian Institution of Washington, repeated Tsiolkovsky's calculations, being unaware of Tsiolkovsky's work at the time. Four years later, the German professor Hermann Oberth drew public attention to rocketry in his famous book, *Die Rakete zu dem Planetenraum,* in which he discussed the use of rocket vehicles for interplanetary travel. It was this work that first stimulated genuine public interest in the possibility of space travel. In 1946, a group of engineers at the Douglas Aircraft Company made the first detailed calculations of what would have to be done to put an artificial satellite in Earth orbit. A little more than a decade later, the Russians achieved this objective by using a converted military rocket (the SS-6) to launch the world's first man-made satellite on October 4, 1957. Thus, the dreams of Tsiolkovsky, Oberth, and Robert Goddard were fulfilled.

The rocket is a reaction motor that generates thrust (or a force) that can propel the rocket by the principle stated in Newton's third law of motion (see Chapter 1). Figure 5.1 shows a schematic diagram of a rocket. It consists of a fuel tank, an oxidizer tank, a rocket engine or reaction motor, and, of course, a payload. The oxidizer is necessary because the rocket is designed to fly outside Earth's atmosphere, where it is not possible to use atmospheric oxygen to burn the fuel. The rocket motor consists of a combustion chamber in which the fuel is burned and a nozzle that creates a stream of high-speed gas that provides the thrust to move the rocket.

The "rocket equation" can now be derived by looking at the conservation of linear momentum in the coordinate system of a rocket moving in free space:

$$M_R(t)\, dV_R = (-dM_R)v_E \tag{5.1}$$

In equation (5.1), $M_R(t)$ is the mass of the rocket. It will be a function of time because the fuel is being consumed as the rocket moves so that its mass will be decreasing until the fuel is exhausted, and dV_R is the corresponding velocity change of the rocket. The left side of equation (5.1) is the change in momentum of the rocket, or the "action" in Newton's law.

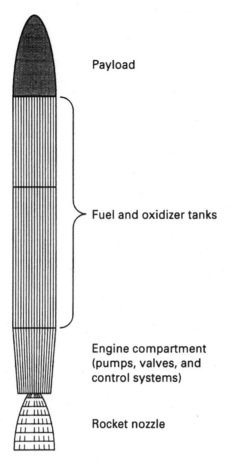

Payload

Fuel and oxidizer tanks

Engine compartment
(pumps, valves, and
control systems)

Rocket nozzle

FIGURE 5.1 Schematic of a single-stage rocket. The major components are the
payload, the fuel and oxidizer tanks, and the rocket engine.

The right-hand side of equation (5.1) is the "reaction." It represents the
decrement in mass of the rocket (hence, the negative sign) as the fuel is
burned and v_E the exhaust velocity of the gas as it leaves the nozzle measured relative to the rocket.

Equation (5.1) is a simple differential equation that can be rewritten as
follows:

$$\frac{dM_R}{M_R} = -\frac{dV_R}{v_E} \qquad (5.2)$$

Integrating this equation yields

$$\int_{M_R(\text{initial})}^{M_R(\text{final})} \frac{dM_R}{M_R} = -\frac{1}{V_E} \int_{V_R(\text{initial})}^{V_R(\text{final})} dV_R \tag{5.3}$$

and performing the indicated operations yields

$$[\log M_R]_{M_R(\text{initial})}^{M_R(\text{final})} = -\frac{V_R(\text{final})}{v_E} \tag{5.4}$$

Equation (5.4) can be evaluated by making some assumptions. Let the initial velocity of the rocket, v_R, be equal to zero so that the right-hand term of (5.4) is equal to $-V_R/v_E$. The initial and final masses of the rocket on the left side of the equation can be written as

$$M_R(\text{initial}) = M_F(\text{fuel/oxidizer}) + M_R(\text{dry}) + M_P \tag{5.5}$$

and

$$M_R(\text{final}) = M_R(\text{dry}) + M_P \tag{5.6}$$

where in equations (5.5) and (5.6) M_P is the mass of the payload carried by the rocket. Here, $M_R(\text{dry})$ is the weight of the rocket and the empty fuel tanks. Therefore, the final velocity of the rocket under the conditions outlined is

$$V_R = v_E \log \left[\frac{M_F(\text{fuel/oxidizer}) + M_R(\text{dry}) + M_P}{M_R(\text{dry}) + M_P} \right] \tag{5.7}$$

Equation (5.7) is the basic rocket equation. The principal point of interest is that when the argument of the logarithm on the right side is equal to e (approximately 2.718), the velocity of the rocket is equal to the exhaust velocity v_E. Obviously, if enough fuel and oxidizer are on board the rocket, the argument can be larger than e, and therefore, the velocity of the rocket can exceed the exhaust velocity of the propellant gases. The argument of the logarithmic term is usually called the *mass ratio,* and the velocity of the rocket can go as high as the mass ratio permits. This is the essential reason why rockets are good for space travel, and this is the point that Tsiolkovsky first recognized in 1903.

Rockets intended to put satellites in Earth orbit or to launch spacecraft to other bodies in the solar system are launched from the surface of Earth. In that case, the basic differential equation (5.1) has to be modified. One factor that must be included is the effect of gravity and the second is the drag on the rocket vehicle caused by the atmosphere. Calculating the atmospheric resistance is complicated because it depends on the details of the shape and design of the rocket. However, the effect of gravity can be calculated easily, and from a practical viewpoint, it is actually more important. Once again, we use Newton's third law of motion, but we now have to include the action caused by Earth's gravity in the equation. The "action variable" is defined as the change in momentum of the object, but by Newton's second law, this can also be expressed as the force acting on the object multiplied by the time interval over which it acts. Therefore, equation (5.1) can be rewritten as

$$M_R(t)\, dV_R = (-dM_R)v_E - M_R(t)g\, dt \qquad (5.8)$$

where the new term on the right contains the gravitational constant g that determines the force of gravity on the rocket. This equation is considerably more complicated than (5.1) if the gravitational constant g is accurately represented as a function of altitude. However, it is useful to treat g as a constant, which is a reasonable approximation as long as the rocket is close to the ground.

Equation (5.8) can be rewritten as a simple differential equation:

$$\frac{dV_R}{v_E} = -\frac{dM_R(t)}{M_R(t)} - \frac{g}{v_E}dt \qquad (5.9)$$

At the beginning of the flight of the rocket, it can be seen that before the rocket starts lifting off the ground, the first term on the right side of the equation must be larger than the second. The important parameter that determines what happens is the constant g/v_E that appears in the second term on the right side. This constant is usually inverted, and it is called the *specific impulse* of the rocket engine:

$$\tau_s = \frac{v_E}{g} \qquad (5.10)$$

The specific impulse has the dimension of time, and it depends only on the exhaust velocity of the rocket. The exhaust velocity in turn depends on thermodynamic considerations that characterize what happens in the combustion chamber of the rocket and the design of the rocket nozzle. In terms of the thermodynamics, the molecular velocity inside the rocket chamber operating at the temperature T is given as

$$v \approx \sqrt{\frac{3kT}{m}} \qquad (5.11)$$

where k is Boltzmann's constant and m is the molecular weight of the combustion products. The expression (5.11) defines approximately the velocity of sound inside the combustion chamber as well. At room temperature, approximately 300 K, the sound velocity is about 0.3 km/s. In a good combustion chamber, the temperature is something like 4000 K so that the sound velocity is about 1000 m/s. If the rocket works at a relatively low exit pressure, then in terms of exhaust velocity there is an increase of another factor of $\sqrt{2\gamma/(\gamma - 1)}$, where the quantity γ is the ratio of the specific heat of the gas at constant pressure to that at constant volume, that must be considered. If the rocket nozzle is well designed, then the exhaust velocity is multiplied by the factor $\sqrt{2\gamma/(\gamma - 1)}$. For the gas in the combustion chamber, the quantity γ is approximately 1.3. Thus, a good estimate of the exhaust velocity of a well-designed rocket is

$$v_E \approx 3000 \text{ m/s}$$

where a molecular weight of the order of 30 has been assumed. This translates into a specific impulse of about

$$\tau_s \approx \frac{3000}{10} \approx 300 \text{ s} \qquad (5.12)$$

The use of liquid hydrogen as a fuel and liquid oxygen as the oxidizer improves things by about a factor of the $\sqrt{1.67}$ since the molecular weight of the exhaust gas (steam) is 18 rather than 30. Thus, the specific impulse that can be reached is

$$\tau_s \approx 390 \text{ s} \qquad (5.13)$$

It is, of course, this point that makes it so advantageous to use liquid hydrogen as a rocket fuel in spite of the difficulties encountered in handling the material.

In order to see what happens at the end of the rocket's flight trajectory, it is necessary to integrate equation (5.9):

$$\frac{1}{v_E} \int_0^{V_R(\text{final})} dV_R = -\int_{M_R(\text{initial})}^{M_R(\text{final})} \frac{dM_R}{M_R} - \frac{1}{\tau_s} \int_0^{\tau_B} dt \qquad (5.14)$$

In this equation, the quantity τ_B is the *burnout time* of the rocket, which is how long it takes the engine to burn up all the fuel. Performing the indicated operation on equation (5.14) yields

$$\frac{V_R(\text{final})}{v_E} = \log \frac{[M_R + M_P + M_F(\text{fuel/oxidizer})]}{[M_R + M_P]} - \frac{\tau_B}{\tau_S} \qquad (5.15)$$

This equation does not have a "closed" solution and so an iterative process is necessary to understand what it says.

To reach Earth orbit, the final velocity of the rocket must be about 27,750 km/h. If we assume that the specific impulse of the rocket is about 300 s, then the exhaust velocity v_E is about 3 km/s so that the ratio of the two velocities is

$$\frac{V_R(\text{final})}{v_E} = \frac{27,750}{3 \times 3600} = 2.57$$

For a single-stage rocket moving a payload to Earth orbit, the burnout time of the fuel would be approximately 10 min, or 600 s, so that

$$\frac{\tau_B}{\tau_s} = 2.00 \qquad (5.16)$$

Using these results, the logarithmic term in equation (5.15) becomes

$$\log \frac{[M_R + M_P + M_F(\text{fuel/oxidizer})]}{[M_R + M_P]} = 4.57 \qquad (5.17)$$

Inverting the logarithm yields

$$1 + \frac{M_F(\text{fuel/oxidizer})}{[M_R + M_P]} = 96.6 \qquad (5.18)$$

so that we have approximately

$$\frac{M_F(\text{fuel/oxidizer})}{[M_R + M_P]} = 95.6 \qquad (5.19)$$

since unity is small compared to the ratio of the masses in the equation. The conclusion of the calculation, therefore, is that it takes about 96 kgm of fuel to place 1 kgm of stuff—in this case the mass of the rocket plus the payload mass—in Earth orbit. This ratio is actually somewhat larger than that which is typically encountered. The take-off weight of the *Saturn V* rocket was about 3.2 million kgm, and it was capable of putting about 100,000 kgm in Earth orbit. Thus, in the case of the *Saturn V,* it took about 32 kgm of fuel to put 1 kgm of material in Earth orbit.

The reason why the Saturn rocket was more effective than the hypothetical rocket described by equation (5.19) is that it was a staged rocket. In order to reach Earth orbit, the Saturn used two separate stages, the *Saturn I,* a gasoline-liquid oxygen fueled first stage, and the *Saturn IIC,* a liquid hydrogen-liquid oxygen fueled rocket. To understand why "staging" a rocket helps, one needs only to consider the point that in a staged rocket the empty fuel tanks are discarded so that more mass becomes available to be placed in Earth orbit. The most transparent way to see how this works is to look at the mathematics of a staged rocket. Let us start by looking at equation (5.17). The left side of the equation is often called the mass ratio, which in our case is the ratio of the mass placed in Earth orbit and the lift-off mass of the rocket:

$$\frac{[M_R + M_P + M_F(\text{fuel/oxidizer})]}{[M_R + M_P]} = \lambda \qquad (5.20)$$

To develop the equation that governs an "ideal" multistage rocket, let us assume that we have a two-stage vehicle where the mass ratio of the first stage is λ_1 and the mass ratio of the second is λ_2. Also, let us assume that the burnout time of the first stage is τ_B (stage 1) and for the second stage, it is τ_B (stage 2). We can then write the following equations by looking at equation (5.15):

$$\frac{V_R(\text{final} - \text{stage 1})}{v_E} = \log \lambda_1 - \frac{\tau_B(\text{stage 1})}{\tau_s} \qquad (5.21)$$

and

$$\frac{V_R(\text{final} - \text{stage 2})}{v_E} = \frac{V_R(\text{final} - \text{stage 1})}{v_E} + \log \lambda_2 - \frac{\tau_B(\text{stage 2})}{\tau_s} \qquad (5.22)$$

In equation (5.22), we have made the assumption that the exhaust veloci-ty v_E and therefore the specific impulse τ_s are the same for each stage. We will also assume for the sake of simplicity that

$$\lambda_1 \approx \lambda_2 \tag{5.23}$$

which is normally not a very good assumption. Also, we will assume, again for the sake of simplicity, that the burnout time of the first stage is equal to that of the second:

$$\tau_B(\text{stage } 1) = \tau_B(\text{stage } 2) \tag{5.24}$$

Therefore, combining equations (5.21)–(5.24), we have

$$\frac{V_R(\text{final} - \text{stage } 2)}{v_E} = 2 \log \lambda - 2\frac{\tau_B}{\tau_s} \tag{5.25}$$

The only difference between equation (5.25) and equation (5.15) is the factor of 2 on the right-hand side. Solving for the logarithmic term yields

$$\log \lambda = \log\frac{M_R + M_P + M_F(\text{fuel/oxidizer})}{M_R + M_P} = \frac{V_R(\text{final} - \text{stage } 2)}{2v_E} + \frac{\tau_B}{\tau_s} \tag{5.26}$$

The second term in equation (5.26) is identical to that in equation (5.15). The first term is smaller by a factor of 2. Thus, we have, for the logarithm of the mass ratio,

$$\log \lambda = 3.32 \tag{5.27}$$

Inverting the logarithm and following the same procedures that were used to obtain equation (5.19), we get

$$\frac{M_F(\text{fuel/oxidizer})}{[M_R + M_P]} = 28 \tag{5.28}$$

which is not much different from the capability of the first two stages of the *Saturn V* vehicle. What it says is that with a two-stage rocket consid-ered here, it takes about 28 kgm of fuel to lift 1 kgm of things into Earth orbit with a two-stage rocket. This calculation illustrates that, even with the very restrictive assumptions that have been made, there is an enormous advantage achieved by staging.

So far, we have restricted all of our arguments to reaching Earth orbit. It is instructive to conclude this chapter with a short description of how the trip to the Moon was made by the Apollo astronauts. Figure 5.2 shows a photograph of the Apollo *Saturn V* launch vehicle. There are actually four separate stages illustrated that were used to propel the Apollo spacecraft first to the Moon and then back to Earth. The first two stages, *Saturn I* and *Saturn IIC,* were used to place a large (100,000-kgm) payload in Earth orbit, consisting of the *Saturn IVB,* the lunar landing module, the service module, and the command module. Once in Earth orbit, the *Saturn IVC* propulsion unit was employed to place the lunar landing module, the service module, and the command module on a trajectory going to the Moon. Doing this required an additional velocity increment, usually called Δv, over that which was required to place the original spacecraft in Earth orbit. This velocity increment is approximately equal to the difference between the velocity that must be imparted to a spacecraft to put it into Earth orbit and to escape from Earth altogether. Referring back to equation (2.9) in Chapter 2, we have, for the orbital velocity,

$$v(\text{orbital}) = \sqrt{\frac{GM_E}{r_E}} \tag{5.29}$$

for a vehicle moving in an orbit with a radius r, where r is the sum of the Earth radius R_E and the altitude of the satellite, h, above the surface of Earth. The escape velocity can be calculated from the total energy equation (2.32) in Chapter 2:

$$E_r = \tfrac{1}{2}mv^2 - \frac{GmM_E}{r} \tag{5.30}$$

For a vehicle to "escape" from the orbit with radius r, we need to move the vehicle away from Earth orbit to a point infinitely far away from Earth. At that point, the velocity of the vehicle should be zero to minimize the energy necessary to escape to infinity. Thus, the total energy at that point should be zero so that

$$\tfrac{1}{2}v^2(\text{escape}) = \frac{GM_E}{r} \tag{5.31}$$

FIGURE 5.2 *Saturn V* moon rocket on launch pad. Two stages, the *Saturn I* and the *Saturn IIC,* are necessary to place the "moonship" into a trajectory going to the Moon. The moonship consists of the lunar module, which is designed to land on the Moon; the service module, which contains the power supply and the rocket engine that provides the propulsion for the return trip; and finally, the command module, in which the three crew members ride. (Photograph courtesy of NASA.)

or

$$v(\text{escape}) = \sqrt{\frac{2GM_E}{r}} \tag{5.32}$$

The escape velocity from the orbit is thus a factor of $\sqrt{2}$ larger than the orbital velocity, or 39,239 km/h. The Moon is far enough away from Earth so that it is not a bad approximation to use the escape velocity to calculate the required velocity increment to leave Earth orbit and travel to the Moon:

$$\Delta v = 39{,}239 - 27{,}750 = 11{,}489 \text{ km/h}$$

This velocity increment is provided by the third stage of the *Saturn V,* the *Saturn IVB.*

When the *Saturn IVB* burn is completed, the lunar module, the service module, and the command module are on their way to the Moon. When this spacecraft arrives at the Moon, the engine in the service module executes a retroburn, which puts the spacecraft in orbit around the Moon. The lunar excursion module is then separated, and it descends to the lunar surface. When the operations on the lunar surface are complete, the ascent module leaves the Moon and returns to lunar orbit to unite with the command and the service modules. After crew transfer, the ascent module is jettisoned. The motor on the service module is then used to propel the spacecraft on a trajectory back to Earth. Finally, the service module is jettisoned, and the command module returns to Earth. The entire Apollo mission profile is shown in Figure 5.3. The mission profile is defined in terms of the velocity increments that the various rocket motors produce. This is quite natural since the rocket equations derived in this chapter yield velocities and velocity increments. We will return to this subject when we consider orbital transfers in detail in Chapter 10.

A number of good books and articles are available for further reading on the subject of rocket propulsion. Robert H. Goddard's "A Method of Reaching High Altitudes," which was published in 1919 by the Smithsonian Institution, is a seminal work in the field. Subsequently in 1920, Goddard submitted another paper on the same subject, in which he first mentions going into space. Unlike the work of Tsiolkovsky, Goddard's papers describe extensive experimental work that he performed with solid and liquid fueled rockets. Goddard later implemented the work described in these papers by flying the first successful liquid fueled rocket in March 1926 and the first successful gyrostabilized rocket in 1935. The first

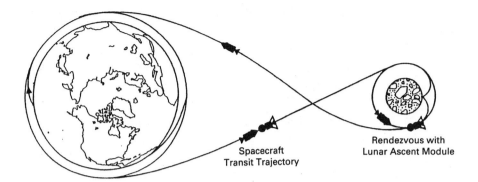

Spacecraft
Transit Trajectory

Rendezvous with
Lunar Ascent Module

Lunar Orbit Rendezvous

FIGURE 5.3 Mission profile used in the Apollo missions to put people on the Moon. It is called the lunar orbit rendezvous method because it involves sending the lunar landing module to the surface of the Moon from the service module and the command module in lunar orbit.

book to draw public attention to the possibility of using rockets to make trips to the planets in the solar system was Hermann Oberth's *Die Rakete zu dem Planetenraum* (*The Rocket to Interplanetary Space*). It was first published in 1923 with an expanded edition following in 1929. In 1936, P. E. Cleator published a comprehensive description of rocket research in the 15 years following 1920, titled *Rockets through Space*. A good history of rocket developments from World War II to the Apollo program in the 1960s is provided in *The Rocket Team* by Frederick I. Ordway and Mitchell R. Sharpe. For an elementary treatment of the technical aspects of rocket propulsion, see Eric Burgess' book *Rocket Propulsion* published in 1954. A more modern technical work is *Rocket Propulsion Elements: An Introduction to the Engineering of Rockets* by George P. Sutton, the sixth edition having been published in 1992.

EXAMPLES

5.1. An interesting variation of the principles of rocket propulsion developed in this chapter is the ion rocket. Instead of using thermal energy to produce the gas stream that propels the rocket, an ion rocket uses charged ions that have been accelerated by an electric field in the rocket motor. The ions are then ejected from the rocket, and they impart a momentum to the vehicle in exactly the same way as described by equation (5.1). There are two principal differences between ion rockets and conventional ones:

1. The exhaust velocity is much higher in an ion rocket; therefore, the specific impulse as defined in equation (5.10) is much larger than for a conventional rocket.
2. The mass ejected by an ion rocket per unit time is much smaller than in the case of a conventional rocket. This happens because it is impossible to achieve the high particle densities (i.e., pressure) in an electrostatic accelerator that is characteristic of the combustion chamber of a conventional rocket.

From these points, it follows that an ion rocket is more "efficient" than a conventional rocket in the sense that it takes a much smaller ejected mass to achieve a given momentum transfer to the vehicle. At the same time, the thrust (i.e., force) developed by an ion rocket is very small because of the low particle density in an ion rocket exhaust stream compared to one in a conventional rocket. The low particle density is caused by the fact that ions are charged particles that repel each other, unlike the neutral atoms in an ordinary gas. This repulsive force between the particles makes it impossible to achieve high particle densities.

It is of interest to quantify the statements that have been made. During the 1970s several orbital flight tests of ion propulsion systems were made, the last one being the SERT II flights from 1979 to 1981. Thus, actual flight data are available on which these calculations can be based. Typically, the ion rocket uses argon as a "working fluid," and the rocket has an electrostatic accelerating system that operates at about 5 kV. A reasonable solar power supply could deliver something like 100 kW to operate the ion-accelerating system, which means that an electric current of about 20 A (as an upper limit) could be produced in the particle beam. [*Note:* Power (watts) equals voltage (volts) times current (amperes).] From these assumptions, it is possible to calculate both the specific impulse of the ion rocket and the thrust that it can develop. The velocity of an argon ion that has been accelerated through a potential difference of 5 kV is given by the equation

$$1/2mv^2 = eV$$

where m is the mass of the argon ion, e the charge on the ion (we shall assume that the ion is singly charged), and V the potential difference through which the ion passes. Thus,

$$v = \sqrt{\frac{2eV}{m}}$$

and, to evaluate the velocity of the ion, we shall use 1.60×10^{-19} C for the charge e, and the mass of the argon atom is approximately 40 times the mass of a hydrogen atom, or 6.68×10^{26} kgm. Therefore,

$$v = 1.55 \times 10^5 \text{ m/s}$$

The specific impulse of this ion rocket is therefore

$$\tau = \frac{1.55 \times 10^5 \text{ m/s}}{10 \text{ m/s}^2} = 15{,}500 \text{ s}$$

This is a factor of 43 better than what can be achieved by the very best liquid-hydrogen oxygen rockets.

In the case of the *Saturn V* vehicle that is discussed in the text, the vehicle weighs about 3.5×10^6 kg. The rate at which mass is ejected varies from stage to stage. The first stage, the *Saturn I*, which is used at lift-off, weighs about 3.0×10^6 kgm, and almost all of that weight is fuel. The burnout time for the *Saturn I* stage was about 5 min, or 300 s. Thus, about 11,700 kgm/s (11.7 tons/s) of mass are ejected by the five F-1 engines of the *Saturn I* stage.

In the case of the ion rocket, the rate at which mass is ejected is calculated by considering the current of the ion beam. Based on available power, we estimated earlier an ion beam current of 20 A. Actually, this is much too high, because we have assumed that the electrical conversion system is 100% efficient. A better estimate for the current would be about 2 A, which assumes an overall conversion efficiency of about 10%. We can now use the fact that each ion carries a charge of 1.60×10^{-19} C and the relationship that

$$1 \text{ A} = 1 \text{ C/s}$$

Using these, we can calculate the number of argon ions that are ejected per second:

$$N = \frac{20}{1.60 \times 10^{-19}} = 1.25 \times 10^{20}$$

Since each argon ion weighs 6.68×10^{-26} kgm, the mass ejected by the ion rocket is approximately

$$\frac{\Delta M}{\Delta t} = 8.36 \times 10^{-6} \text{ kgm/s}$$

Note that this quantity is more than one billion times smaller than the mass ejected by the *Saturn I* vehicle. The ratio of the thrusts of the *Saturn I* and the ion rocket can now be estimated by using the relationship

$$\text{Thrust} = \frac{\Delta P}{\Delta t} = v_E \frac{\Delta M}{\Delta t}$$

where ΔP is the momentum imparted to the vehicle by the exhaust stream. Thus

$$\frac{\text{Thrust(Saturn)}}{\text{Thrust(ion rocket)}} = \frac{v_E(\text{Saturn})}{v_E(\text{ion rocket})} \frac{\Delta M(\text{Saturn})}{\Delta M(\text{ion rocket})} = 2.7 \times 10^6$$

The Saturn rocket, therefore, has nearly three million times the thrust of the ion rocket. From these considerations, it can be concluded that ion rockets are not suited to be launch vehicles for which high thrust is essential. Ion rockets are important if not much thrust is required but if a very high propulsive efficiency is necessary. Examples of this might be long-term station keeping for satellites in Earth orbit or journeys to other planets for which a low-thrust rocket that operates for a long period of time is required.

5.2. Ion rockets are a working proposition, and there is good reason to believe that they will find applications in the not-too-distant future. Another rocket concept that has received considerable attention is the nuclear rocket. In this case, a nuclear reactor is used as the energy source and is operated roughly as shown in Figure 5.4. The working fluid for the rocket would be hydrogen, which would be heated

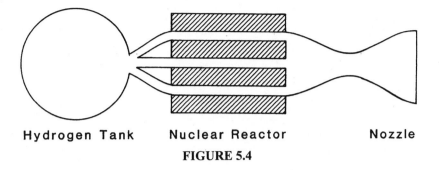

Hydrogen Tank   **Nuclear Reactor**   **Nozzle**

FIGURE 5.4

by passing through a series of channels in a nuclear reactor. Once the hot gas leaves the reactor, it would be passed through an efficient aerodynamic expansion nozzle and ejected. The principal advantage of a nuclear rocket is not that it could operate at higher temperatures but rather that the molecular weight of the ejected material is as small as it can get. With respect to temperature, all rockets are limited to about 4000°C because that is all that materials can stand. However, the specific impulse varies inversely as the square root of the molecular weight [see equations (5.10) and (5.11)], and since hydrogen, which has a molecular weight of 2 rather than 18 for water, can be used as a working fluid, nuclear rockets should be able to achieve specific impulses of the order of 1000 s. Another positive point is that nuclear rockets do not suffer from the severe thrust limitations of ion rockets. Since the particles in the working fluid are not charged, high densities and large mass flows can be maintained.

Much experimental work has been performed with nuclear rockets. A full-scale nuclear rocket (the NOVA) was built at the Los Alamos National Laboratory, and it was operated a number of times successfully during the 1970s. There is no doubt that a technically successful nuclear rocket can be built. The trouble is that there are many safety problems to solve and political problems to overcome before such a system could be flown. At the present time, work on nuclear rockets has been suspended.

5.3. Several other much more exotic rocket concepts have been considered. One would be to use electrons rather than ions as a working fluid in a rocket. The principle would be the same as the ion rocket that has been described. The electron is 1830 times lighter than a proton and 73,200 times lighter than the argon atoms used in the first example. Thus a significant advantage in specific impulse would be gained. The principal problem is that electron beams would be much harder to maintain and to keep properly focused than ion beams at the same energy.

A second concept is laser propulsion: In this case, a laser beam would be focused on a working fluid tank located at the bottom of the rocket. The energy in the beam would heat the fluid, and the fluid would then be ejected through an appropriate nozzle to provide thrust. As the rocket rises through the atmosphere, the laser beam would continue to follow the vehicle and would continue to heat the working fluid tank. The putative advantage of this scheme is that the energy source for the rocket would stay on the ground so that the weight devoted to carrying the energy source would be saved. No practical work has been done on this concept.

Finally, there is the "Orion" concept attributed to Freeman Dyson. This idea involves nuclear explosives used to drive an appropriately designed space ship. The ship would be fitted with a shock-absorbing plate at one end against which the nuclear blast would work. Very high speeds on long voyages could be achieved by releasing successive explosive devices and detonating them behind the shock-absorbing plate. No practical work has been done on this concept.

PROBLEMS

5.1. Alpha Centauri is 4 light years away from Earth. An ion rocket has been built to reach the star. The velocity of the ions emitted by the ion rocket is one-tenth of the speed of light. The ratio of the initial mass of the rocket vehicle to the final mass of the rocket vehicle is 3, and it takes about 100 h to exhaust all of the working fluid of the ion rocket. How long does it take the vehicle to reach Alpha Centauri? Neglect effects of relativity since the ion velocity is small enough compared to the velocity of light.

5.2. An ion rocket is designed so that the exhaust velocity is one-tenth the velocity of light. We want to compare its performance to a conventional rocket with an exhaust velocity of 1.86 miles per second. For the ion rocket, 1 pound of fuel accelerates 1 pound of payload to a final velocity v. What is the mass of fuel necessary to accelerate the same payload to the same final velocity with the conventional rocket? Neglect the dry vehicle weight in each case compared to the payload weight.

5.3. Rockets A and B both carry a payload of mass m. The fuel mass of each rocket is M_F and the empty mass of each rocket is M_R. The total mass of each is M. Rocket A has one stage and rocket B has two stages each with half the fuel and half the empty weight of rocket A (see Figure 5.5). The exhaust velocity of all the rocket engines is v_E. Both rockets are at rest in free space when the engine is started.

(a) Derive the formula for the final velocity of the payload for rocket A.

(b) Derive the formula for the final velocity of the payload for rocket B.

(c) Discuss the significance of the results of parts 1 and 2.

5.4. Sounding rockets are designed to explore the upper atmosphere and therefore do not possess the ability to reach Earth orbit. A solid fueled sounding rocket weighing 1000 pounds is launched vertically from the ground. The payload carried is 10 pounds and the weights of the rocket casing and the nozzle are so small that they can be ne-

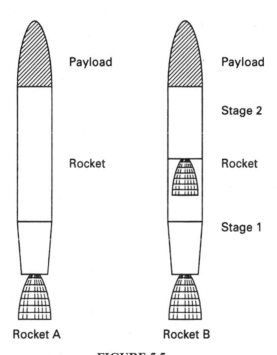

FIGURE 5.5

glected compared to the weight of the payload. The exhaust velocity of the rocket, v_E, is 4000 ft/s, and the drag of the air and the rotation of Earth can both be neglected. The solid fuel of the rocket burns out after 100 s.

(a) What is the final velocity of the rocket at burnout?

(b) At what altitude does the rocket burn out? Assume that the rocket fuel burns at constant rate and that the ratio of the payload mass to the initial fuel mass can be set equal to zero in your calculations.

(c) What is the highest altitude reached by the rocket? Assume that Earth's force of gravity is constant, since the sounding rocket is always close to Earth.

(d) What is the velocity of the payload when it returns to the ground?

CHAPTER 6

ENERGY RELATIONSHIPS: HYPERBOLIC AND PARABOLIC ORBITS

In the preceding chapters, we examined the behavior of two masses moving around each other under the influence of the gravitational force between them. In Chapter 4, we discussed the behavior of elliptic orbits in some detail. We looked at the geometric properties of the ellipse and then proved Kepler's three laws of planetary motion by also treating the dynamics of objects moving in elliptic orbits. Finally, we derived the relationship between the kinetic and the potential energy of an object moving in a circular orbit. The purpose of this chapter is to extend all of these considerations to other types of two-body orbits permitted by the equations of motion and the law of gravity. These orbits are the parabola and the hyperbola.

We shall start by looking at the energy relationships in an elliptic orbit. We have already shown that the total energy of two objects moving under the influence of gravity is constant; see equation (3.42) in Chapter 3. We have also shown that in the case of circular orbits the kinetic energy and the potential energy are both constants because the orbital speed and the radius are both constant. Using the expression for the orbital speed in a circular orbit (2.9), it was possible to write the total energy of a mass m moving around a mass M at a radius R as

$$E_T = -\frac{GmM}{2R} \tag{6.1}$$

We shall now extend these considerations to elliptic orbits and calculate the total energy of the mass m moving around the mass M in an orbit, such as the one shown in Figure 6.1. The total energy of the mass m can be written, according to equation (3.42), as

$$E_T = \tfrac{1}{2} mv^2 - \frac{GmM}{r} \tag{6.2}$$

In this case, the kinetic energy and the potential energy are no longer constant; they change as the mass moves around the orbit. In order to calculate the total energy of the particle in the orbit, we use the conservation of angular momentum in the orbit. To do this, we remember that the angular momentum is given by equation (3.47):

$$\mathbf{L} = m\mathbf{r} \times \mathbf{v} \tag{6.3}$$

When the vectors $\mathbf{r}$ and $\mathbf{v}$ are perpendicular to each other, the magnitude of the angular momentum vector $\mathbf{L}$ is given as

$$\mathbf{L} = |\mathbf{L}| = mrv \tag{6.4}$$

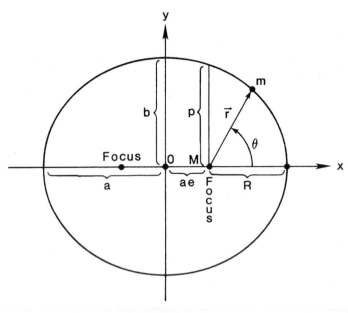

FIGURE 6.1

There are only two places along the elliptic orbit where $\mathbf{v}$ and $\mathbf{r}$ are perpendicular, at perigee and at apogee. At apogee, we have

$$L = mv_A(a + ae) = mv_A a(1 + e) \tag{6.5}$$

Since

$$a = \frac{p}{1 - e^2} \tag{6.6}$$

we have

$$L = mv_A \frac{p}{1 - e} \tag{6.7}$$

The angular momentum is related to the semilatus rectum p of the ellipse as follows:

$$p = \frac{L^2}{GMm^2} \tag{6.8}$$

From the relationships (6.6)–(6.8), we can calculate v_A, the velocity at apogee:

$$v_A^2 = \frac{L^2}{m^2} \frac{(1 - e)^2}{p^2} = \frac{GM(1 - e)^2}{p} = \frac{GM}{a} \frac{1 - e}{1 + e} \tag{6.9}$$

Using equation (6.2), we can write, for the total energy at apogee,

$$E_T = \frac{1}{2}mv_A^2 - \frac{GmM}{r_A} = \frac{1}{2}\frac{GmM}{a}\frac{1 - e}{1 + e} - \frac{GmM}{a(1 + e)} \tag{6.10}$$

where the relationship $r_A = a(1 + e)$ has been employed. When equation (6.10) is simplified, we obtain

$$E_T = -\frac{GmM}{2a} \tag{6.11}$$

The total energy of the particle, m, moving along the elliptic orbit is a constant that depends only on the semimajor axis of the ellipse. The total energy is always negative, and it has exactly the same functional form as equation (2.35), the total energy of a particle moving in a circular orbit, except that r, the radius of the circle, is replaced by a, the semimajor axis.

The equation of an ellipse in polar coordinates was derived in Chapter 4 [see equation (4.64)], and it looks like this:

$$r = \frac{p}{1 + e \cos f} = \frac{p}{1 + e \cos \theta} \tag{6.12}$$

where we have assumed that the angle θ_0 defined in Chapter 4 is zero. If the eccentricity e of the orbit is greater than zero but less than unity, then r remains finite, no matter what the value of the angle θ. Such orbits are called "bounded" since they can only occupy a finite space. The fact that the total energy for bounded orbits (ellipses and circles) turns out to be always negative is not surprising. If the body in orbit (mass m) were to be moved from its orbit to a point infinitely far away from M, and with no kinetic energy, work would have to be done on the body. This work precisely equals the total energy of the body (m) in its orbit, and it would be added to that energy so that, in the end state, the total energy of the body is zero; that is, the kinetic energy vanishes because v_∞ is zero, and the potential energy also vanishes because the distance to the force center becomes infinite.

In case $e = 0$, then equation (6.12) becomes

$$r = p \tag{6.13}$$

and the orbit is a circle, and it is the simplest of orbits. We have now fulfilled the promise made in Chapter 2, and we have shown that a circle is indeed a permitted orbit.

What if $e = 1$? In this case, the equation of the orbit is

$$r = \frac{p}{1 + \cos \theta} \tag{6.14}$$

When the angle θ is equal to zero, we have

$$r = \tfrac{1}{2}p \tag{6.15}$$

On the other hand, if the angle θ is equal to π (180°), then r tends to infinity. Finally, when $\theta = \tfrac{1}{2}\pi$, we have

$$r = p \tag{6.16}$$

Figure 6.2 shows the kind of orbit that is implied by these considerations. To see what this orbit would look like, we look at the transformation introduced in Chapter 4 [equation (4.39)]:

$$x = r \cos \theta \quad y = r \sin \theta \tag{6.17}$$

Substituting these equations into (6.14) gives

$$r + r \cos \theta = p \tag{6.18}$$

or

$$\sqrt{x^2 + y^2} = p - x$$

where we have used equations (6.5). Squaring both sides gives

$$x^2 + y^2 = p^2 - 2xp + x^2 \tag{6.19}$$

which when simplified becomes

$$y^2 = p^2 - 2xp \tag{6.20}$$

If we make the transformation,

$$-x' = x - \tfrac{1}{2}p \tag{6.21}$$

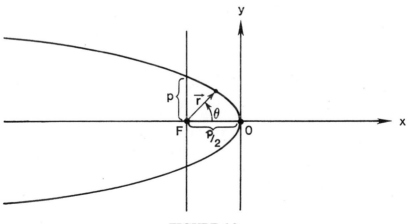

FIGURE 6.2

we obtain

$$x' = \frac{y^2}{2p} \tag{6.22}$$

This is the equation of the parabola shown in Figure 6.2.

Now, we should look at what is implied by an eccentricity equal to unity. One way of looking at this is to say that the parabola we have drawn in Figure 6.1 is the extreme case of an ellipse. Referring back to Figure 4.2 in Chapter 4, we can compare this to the parabola shown in Figure 6.2. We see first that the origin of the Cartesian coordinate system that describes the parabola in equation (6.9) is at what would be the focus of the ellipse in Figure 4.2. Furthermore, from the equations that relate the quantities a and b for the ellipse, we have

$$a = \frac{p}{1 - e^2} \quad \text{and} \quad b = \frac{p}{\sqrt{1 - e^2}} \tag{6.23}$$

It is clear from these equations that both a and b become infinite if $e = 1$. Referring back to Figure 6.2, it is as if the ellipse shown in Figure 6.2 were "stretched" so that both the semimajor axis a and the semiminor axis b move infinitely far to the left. Also, both a and b become infinitely long. Finally, the "center" of the ellipse, point 0, moves infinitely to the left. Since the semimajor axis goes to infinity, the total energy of a body moving in parabolic orbit is zero, according to equation (6.11). If the mass m moves in a parabolic orbit, the dynamic condition that describes the motion is that the potential energy of the mass is always precisely equal to the kinetic energy:

$$\tfrac{1}{2}mv^2 = \frac{GMm}{r} \tag{6.24}$$

Because of the exact nature of the condition in (6.24), it is very difficult to achieve orbits that precisely fit a parabola.

We now turn to a study of the cases in which the eccentricity is greater than unity ($e > 1$). If we look at equation (6.12), the first and most obvious point is that there is an angle θ for which r becomes infinite. If we look at equation (6.12), then there is an angle $\theta = \alpha$ for which r goes to infinity when this condition is fulfilled:

$$1 + e \cos \alpha = 0 \qquad \cos \alpha = -\frac{1}{e} \tag{6.25}$$

Note that the implication of equation (6.25) is that

$$\alpha > \tfrac{1}{2}\pi \qquad\qquad (6.26)$$

because it is only when this condition is fulfilled that the cosine of the angle is negative. Also, when $\theta = 0$, we have

$$r = \frac{p}{1 + e} \qquad\qquad (6.27)$$

and when $\theta = \tfrac{1}{2}\pi$, we have

$$r = p \qquad\qquad (6.28)$$

Conditions (6.27) and (6.28) along with (6.12) now permit us to construct a picture of the orbit (Figure 6.3). This picture shows that there can be no angles larger than α, for which solutions to (6.12) exist. The straight line that makes the angle α with the x axis comes closer and closer to the orbit as r becomes larger and larger. Such a line is called an asymptote, and in the case of the parabola, there is no asymptote.

Now, let us look at the energy equation

$$E_T = \tfrac{1}{2}mv^2 - \frac{GMm}{r} \qquad\qquad (6.29)$$

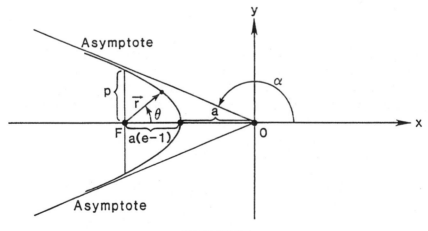

FIGURE 6.3

Note here that the total energy is equal to the kinetic energy when the value of r goes to infinity. In order to determine what the total energy is in this case as given by equation (6.11) for the ellipse, we will again transform the polar coordinates to Cartesian coordinates using

$$x = r \cos \theta \qquad y = r \sin \theta$$

and

$$r^2 = x^2 + y^2$$

From this we can write

$$\sqrt{x^2 + y^2} = \frac{p}{1 + e(x/r)} = \frac{p}{1 + e\,x/\sqrt{x^2 + y^2}}$$

$$\sqrt{x^2 + y^2} = \frac{p\sqrt{x^2 + y^2}}{\sqrt{x^2 + y^2} + ex} \tag{6.30}$$

Simplifying (6.30) leads to

$$\left(\sqrt{x^2 + y^2} + ex\right) = p \tag{6.31a}$$

$$\sqrt{x^2 + y^2} = p - ex \tag{6.31b}$$

Squaring both sides of equation (6.31) yields

$$x^2 + y^2 = p^2 - 2pex + e^2 x^2 \tag{6.32}$$

In order to rewrite equation (6.32) in terms of x^2 and y^2 only, a transformation of coordinates is necessary. Let us look at Figures 6.1 and 6.3 and see which quantity in the case of the orbit in Figure 6.1 is the analog of the semimajor axis of the ellipse. In the case of the ellipse, it is the distance from the center of the ellipse, O, to the point P. There is no natural center of the orbit in Figure 6.3 since the orbit is not closed. A "natural" choice would be to see if the point at which the asymptote crosses the x axis is an appropriate origin.

Look at equation (6.32) again and see what can be done:

$$(1 - e^2)x^2 + y^2 = p^2 - 2pex \tag{6.33}$$

and this can be restated as

$$x^2 + \frac{y^2}{1 - e^2} = \frac{p(p - 2ex)}{1 - e^2}$$

Now, remembering that for the hyperbola $e > 1$ we have

$$x^2 - \frac{y^2}{e^2 - 1} = \frac{p(2ex - p)}{e^2 - 1} \tag{6.34}$$

Now we need to make a transformation of the x coordinate to eliminate the linear term. This can be done by completing the square in (6.34), which leads to the following final form:

$$\left(x - \frac{pe}{e^2 - 1}\right)^2 - \frac{y^2}{e^2 - 1} = \frac{p^2}{e^2 - 1^2} \tag{6.35}$$

From equation (6.35), it is clear what the required transformation should look like:

$$x' = x - \frac{pe}{e^2 - 1} \tag{6.36}$$

So that the equation of the hyperbola is

$$x^2 - \frac{y^2}{e^2 - 1} = \frac{p^2}{e^2 - 1^2} \tag{6.37}$$

or in the format

$$\frac{x'^2}{a^2} - \frac{y^2}{b^2} = 1 \tag{6.38}$$

We have

$$\frac{x'^2}{p^2/(e^2 - 1)^2} - \frac{y^2}{p^2/(e^2 - 1)} = 1 \tag{6.39}$$

Therefore, we have the relationships for the quantities a^2 and b^2 in the standard equation for the hyperbola:

$$a^2 = \frac{p^2}{(e^2 - 1)^2} \quad \text{or} \quad a = \pm\frac{p}{e^2 - 1}$$

$$b^2 = \frac{p^2}{e^2 - 1} \quad \text{or} \quad b = \pm\frac{p}{\sqrt{e^2 - 1}} \tag{6.40}$$

Note that these relations are similar to those derived for the ellipse, equation (6.23). Therefore, the functional form of the energy relationship for hyperbolic orbits is the same as that for the ellipse. There is, however, one critical difference: Unlike elliptic orbits, hyperbolic orbits are "unbounded," that is, the body (mass m) in orbit can move infinitely far away from the force center (mass M), and it can even have a residual velocity at infinity, v_∞. Because the potential energy vanishes as r approaches infinity, the total energy at infinity is therefore

$$E_T = \tfrac{1}{2}mv_\infty^2 \tag{6.41}$$

This quantity is always positive, and since energy is conserved, the total energy of the mass m moving in the hyperbolic orbit is always positive. We have already established that the functional form of the total energy for the hyperbola is the same as for the ellipse. To obtain a positive energy means, mathematically, using the negative rather than the positive square root in equation (6.40) to determine the quantity a. Thus, we have, for the hyperbola,

$$E_T = \frac{GmM}{2a} \tag{6.42}$$

Finally, it is instructive to look at what happens as the eccentricity e of the hyperbola varies from unity to infinity, which is the range of values it can take on. For values of e tending toward unity (e cannot equal unity since we then have a parabola), we obtain approximately

$$\cos \alpha \approx -1$$

so that $\alpha \approx \pi$. This means that the hyperbola degenerates into a straight line along the x axis of the Cartesian coordinate system. Also, a tends toward negative infinity, which means that the total energy tends to zero. Thus, the figure becomes a degenerate parabola. For values of e tending toward infinity, we have

$$\cos \alpha = 0$$

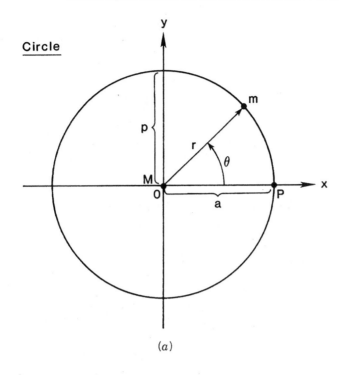

(a)

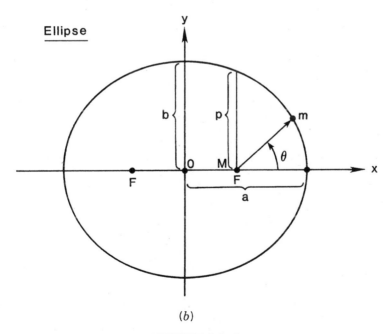

(b)

FIGURE 6.4a–b

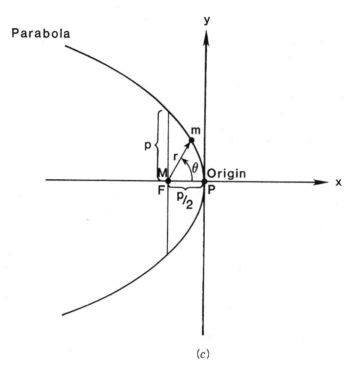

(c)

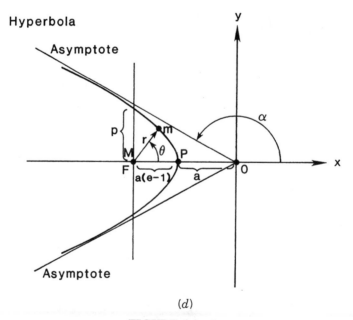

(d)

FIGURE 6.4c–d

which means that $\alpha = \frac{1}{2}\pi$. Also, in this case, a tends toward zero so that the total energy becomes very large, tending toward infinity. Thus, the orbit tends toward a straight line coincident with the y axis of the Cartesian coordinate system.

We now have established all of the relationships outlined in the first paragraph of this chapter.

It might be useful to summarize all of the relationships derived in this chapter with some drawings. It is clear that the circle and parabola are very special cases for which $e = 0$ and $e = 1$, respectively. Ellipses have $0 < e < 1$ and hyperbolas have $e > 1$. Figure 6.4 *a–d* illustrate these points.

For an excellent treatment of the mathematics of conic sections, the curves that are treated in this chapter, see *Calculus* by Tom M. Apostol.

PROBLEMS

6.1. A comet starts from a point at rest that is 10 light years away from the Sun and begins falling in a straight line toward the Sun. How fast is the comet moving when it passes the orbit of Earth? Earth is 8 light minutes away from the Sun. Assume that one year is 365 days and that the speed of light is about 300,000 km/s.

6.2. Halley's comet has a period of 76 years. Its closest approach to the Sun is 0.6 a.u. (An astronomical unit is about 150,000,000 km, the distance from Earth to the Sun.) How far away from the Sun does its orbit extend?

6.3. A comet is observed to be approaching the Sun on a trajectory determined to have a total energy of zero. The comet crosses Earth's orbit, swings around the Sun, and recrosses Earth's orbit at a point precisely on the opposite side of Earth's orbit from the initial crossing.

(a) What is the closest distance of approach of the comet to the Sun?

(b) How much time does the comet spend inside Earth's orbit? The methods developed in Chapter 7 must be used to calculate the answer.

CHAPTER 7

KEPLER'S EQUATION AND LAMBERT'S THEOREM

Up to this point, we have been concerned with the problem of determining the geometric properties of the orbits of bodies moving with respect to each other under the influence of mutual gravitational attraction. We have shown that circles, ellipses, parabolas, and hyperbolas are all permitted orbital trajectories. We have also touched upon dynamics, that is, the behavior of the two-body system when the time variable is introduced, and shown that both the angular momentum and the total energy of the two-body system are constants of the motion. Finally, we have shown that Kepler's laws of planetary motion (see Chapter 1) follow from the law of gravity and Newton's laws of motion. Having developed these results, we can now consider the more difficult and more important problem of determining what kind of orbit is executed by a body moving under the influence of gravity around another one from direct measurements of the position of the body at various times. This is, of course, the problem Kepler first solved in 1609 when, using excruciatingly tedious hand calculations and Tycho Brahe's superb measurements, he established his famous laws.

The problem we want to solve can be stated as follows: Given measurements of the position of a body moving in the gravitational field of another one at various times, how can one determine the orbit? We shall show that if the position of the body, r_1, with respect to some coordinate system is measured at time t_1 and again at r_2 at time t_2, then it is possible to uniquely determine the orbit that the body executes. This is essentially the same as the problem originally solved by Kepler. However, we have much more powerful and elegant techniques today to do so. Remember, Kepler did his work almost 80 years before the invention of differential calculus. There are two relationships that are useful for determining orbits using measurements of the position of an object in orbit at various times. One of these is called Kepler's equation, and the other is a theorem developed in 1761 by J. H. Lambert. The only difference between the two equations that result is essentially the choice of origin of the coordinate system in which the measurements to determine the orbits are made. In both cases, two quantities are to be determined from the measurements. The eccentricity of the elliptic orbit, e, and the semimajor axis a. We shall consider both elliptic and hyperbolic orbits, and we will derive the relationships for both types. Indeed, the methods to be described here were first worked out to find out whether the orbits of various comets that pass close to the Sun were elliptic (closed or bounded) or hyperbolic (unbounded).

7.1 DERIVATION OF KEPLER'S EQUATION

The essential idea behind Kepler's equation is an extension of Kepler's third law, which states that the square of the orbital period of a body moving in an elliptic orbit is proportional to the cube of the semimajor axis (a) of the ellipse. This relationship was derived by integrating the angular momentum equation around the entire elliptic orbit (see Chapter 4). If we are interested in determining what kind of orbit a body executes from measurements that are made, then the angular momentum equations expressed in terms of the orbital parameters are integrated only over a *finite segment* of the orbit. In this way, the desired relationships between the times at which the measurements are made and the position of the body in the orbit can be derived.

In order to accomplish the objective outlined in the previous paragraph, it is useful to look at Figure 7.1. This illustrates another way of constructing an ellipse by enclosing it between two circles, one with a radius a, the semimajor axis, and the other with a radius b, the semiminor axis. The ellipse is then generated by tracing point Q in such a way that the relationship

$$\frac{a}{b} = \frac{CD}{OD} \tag{7.1}$$

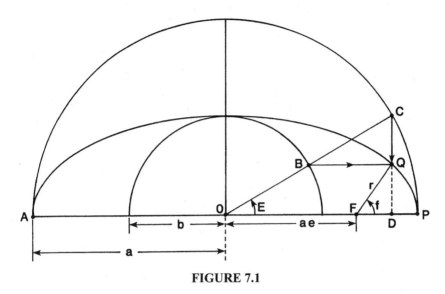

FIGURE 7.1

is always preserved. It is clear from this condition that the line segment BQ is always parallel to the line segment OD. There are two angles that are important in defining the problem being considered. One is the angle θ, which defines the angle that the line segment r makes with the x axis. If the major axis of the ellipse is not colinear with the x axis of the coordinate system, then we must use the relationship defined in Chapter 4, equation (4.63)

$$f = \theta - \theta_0 \tag{7.2}$$

in defining the equation of the ellipse. Here, θ_0 is the angle between the major axis of the ellipse and the x axis of the coordinate system, and the angle f is called the *true anomaly*. The angle E in Figure 7.1 is called the *eccentric anomaly*, and it is the angle between the line OC and the major axis of the ellipse. It turns out that the integration of the angular momentum equation part way around the elliptic orbit yields a particularly simple relationship between the time variable t and the eccentric anomaly E, and it is this relationship that we shall now derive.

We shall start with the equation describing an ellipse in polar coordinates:

$$r = \frac{p}{1 + e\cos(\theta - \theta_0)} = \frac{p}{1 + e\cos f} \tag{7.3}$$

given in equation (4.64) in Chapter 4. In this chapter, we shall use the true anomaly f instead of the polar angle θ in order to preserve the possibility that the major axis of the elliptic orbit may not be colinear with the x axis of the coordinate system used to describe the problem. From the geometry in Figure 7.1, we can write

$$r^2 = (QD)^2 + (FD)^2 \tag{7.4}$$

Also, from Figure 7.1, we have

$$a \cos E = ae + FD$$

or

$$FD = a \cos E - ae \tag{7.5}$$

and

$$QD = b \sin E \tag{7.6}$$

Substituting equations (7.5) and (7.6) into equation (7.4) and performing some algebraic manipulation yields

$$r^2 = b^2\sin^2 E + a^2\cos^2 E - 2a^2e \cos E + a^2e^2$$
$$= b^2\sin^2 E + (a \cos E - ae)^2 \tag{7.7}$$

and then, using the relationship

$$b = a\sqrt{1 - e^2} \tag{7.8}$$

yields, after some more algebraic manipulation,

$$r^2 = a^2(1 - e \cos E)^2$$

or

$$r = a(1 - e \cos E) \tag{7.9}$$

This equation represents the equation of the ellipse shown in Figure 7.1 in terms of the eccentric anomaly E.

We are now ready to establish the relationship between the time variable t and the eccentric anomaly E by using the angular momentum relationship defined in equations (4.11) and (4.63):

$$mr^2 \frac{df}{dt} = L \tag{7.10}$$

where L is the angular momentum, which is a constant of the motion. Recall that when this equation was integrated around the entire orbit of the ellipse (E or θ or f going from 0 to 2π), the resulting relationship was Kepler's third law of planetary motion. What we will now do is to integrate equation (7.10) only part of the way around the orbit in order to derive the desired relationship. Using equation (4.61) of Chapter 4, we will start with the relationship

$$\frac{L^2}{m^2} = GMa(1 - e^2) \tag{7.11}$$

where the masses m and M are located at the points Q and F, respectively, in Figure 7.1. Rewriting equation (7.11) using (7.10), we obtain

$$r^2 \frac{df}{dt} = \sqrt{GMa(1 - e^2)} \tag{7.12}$$

Rewriting equation (7.12) in terms of the eccentric anomaly E and using equation (7.9) give

$$a^2(1 - e \cos E)^2 \frac{df}{dE} \frac{dE}{dt} = \sqrt{GMa(1 - e^2)} \tag{7.13}$$

The integral that needs to be evaluated can now be written as follows:

$$\int_0^{E_0} a^2(1 - e \cos E)^2 \frac{df}{dE} dE = \int_0^{t_0} \sqrt{GMa(1 - e^2)} dt \tag{7.14}$$

As we have already stated, the integrals here are not evaluated around the entire orbit but only to the eccentric anomaly (angle) E_0, which corresponds to the position of the body of mass m at the time t_0. In order to evaluate the integral on the left side of equation (7.14), an expression for df/dE must be obtained. By comparing the equations for the elliptic orbit in terms of the angle f [equation (7.3)] and the angle E [equation (7.9)], the following relationship can be derived:

$$\cos f = \frac{e - \cos E}{e \cos E - 1} \tag{7.15}$$

It is a complicated but straightforward procedure to calculate the desired derivative (df/dE) equation (7.15), and the result is

$$\frac{df}{dE} = \frac{\sqrt{1 - e^2}}{e \cos E - 1} \tag{7.16}$$

Substituting equation (7.16) into the integral (7.14) yields

$$\int_0^{E_0} a^2 (e \cos E - 1) \sqrt{1 - e^2} \, dE = \int_0^{t_0} \sqrt{GMa(1 - e^2)} \, dt \tag{7.17}$$

where the limits of integration are such that the eccentric anomaly is zero at $t = 0$ and E_0 when $t = t_0$. Rearranging the terms and evaluating the integral yields the equation

$$E_0 - e \sin E_0 = \sqrt{\frac{GM}{a^3}} \, t_0 \tag{7.18}$$

Using equation (4.62), which is the statement of Kepler's third law of planetary motion, we can write

$$\sqrt{\frac{GM}{a^3}} = \frac{2\pi}{T} \tag{7.19}$$

where T is the period of the mass m in executing the elliptic orbit shown in Figure 7.1. Therefore,

$$t_0 = \frac{T}{2\pi} (E_0 - e \sin E_0) \tag{7.20}$$

This is called Kepler's equation. If the initial condition is not as chosen, but rather that the time interval for the angle E to change from E_1 to E_2 is $t_2 - t_1$, then equation (7.20) can be rewritten as

$$t_2 - t_1 = \frac{T}{2\pi} [(E_2 - E_1) - e(\sin E_2 - \sin E_1)] \tag{7.21}$$

This equation says that the parameters that describe an elliptic orbit, the eccentricity e, and the semimajor axis a can be uniquely determined by making two measurements. The first is a measurement of the orbital period T, which determines the semimajor axis a. The second is a measurement of the time interval $t_2 - t_1$ that it takes the object in the orbit to move through the angle $E_2 - E_1$. This measurement determines e, the eccentricity of the ellipse.

Kepler's equation is transcendental, and therefore, it has no algebraic solution. To get some physical insight into how the equation works, we will look at the approximation in which the angle E_0 as measured from the major axis of the ellipse shown in Figure 7.1 is very small. In that case, we can use the approximation that

$$E_0 \approx \sin E_0$$

and, therefore, we have, from Kepler's equation (7.20), that

$$E_0 \approx \frac{2\pi t_0}{T(1 - e)} \qquad (7.22)$$

This equation illustrates the point that (again, for small angles, E_0) given a measured time interval t_0, the angle E_0 will be larger for large values of the eccentricity e. This leads to the correct conclusion that the larger the eccentricity of an elliptic orbit, the larger the velocity of the orbiting object at the perigee of the orbit.

The part of Kepler's equation that contains the angle E_0 is often called the *mean anomaly*:

$$\frac{2\pi t_0}{T} = E_0 - e \sin E_0 = l \qquad (7.23)$$

Together with the two other angles, f and E, that have been previously defined as "anomalies," the three anomalies are shown in Table 7.1.

We have seen that in the case of elliptic orbits Kepler's equation is derived starting with the conservation of angular momentum. Angular momentum is also conserved in the case of hyperbolic orbits, so that a version of Kepler's equation for such orbits can also be developed. The geometry is illustrated in Figure 7.2. The left branch of the hyperbola represents the orbit governed by a mass M located at F. The center of the hyperbola is at O and the pericenter at P. The distance between the center

TABLE 7.1

Symbol	Terminology	Description	Radial Distance
f	True anomaly	Angle at focus of the ellipse	$a(1 - e^2)/(1 + e \cos f)$
E	Eccentric anomaly	Angle at center of the ellipse	$a(1 - e \cos E)$
l	Mean anomaly	$(2\pi/T)(t_2 - t_1)$	

and the pericenter is a, which is the semimajor axis. Note the analogy with elliptic orbits as shown in Figures 4.1 and 7.1. The semimajor axis a = PO, and the distance OF = ae for both kinds of orbits. The pericenter distance for a hyperbola is $FP = a(e - 1)$, but for an ellipse it is $a(1 - e)$. The semiminor axis is $b = a\sqrt{e^2 - 1}$ for a hyperbola and $a\sqrt{1 - e^2}$ for an ellipse. Note that for an ellipse the sum of the distances from the foci is 2α, and for a hyperbola, the difference between those distances is 2α. For instance, the distance is $F'P$ is $ae + a = a(e + 1)$, and the distance FP is $a(e - 1)$. The difference is $2a$. Similarly, the length of the semilatus rec-

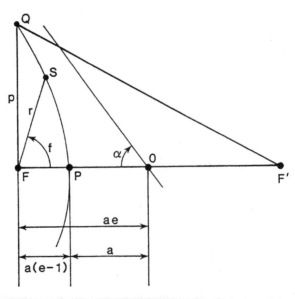

FIGURE 7.2

tum is $FQ = p$. The distance of the point Q from F' is $Qf' = \sqrt{p^2 + (2ae)^2}$. In order to verify the previously given formula for p, we write

$$QF' - QF = 2a$$

or

$$\sqrt{p^2 + 4e^2a^2} - p = 2a$$

from which $p = a(e^2 - 1)$.

The angle of the asymptotes may be obtained from finding the angle f_0 from the equation $1 + e \cos f_0 = 0$, since this corresponds to $r \to \infty$. The solution is

$$f_0 = \arccos\left(\frac{-1}{e}\right)$$

and therefore, $\alpha = \pi - f_0$ or $\alpha = \arccos(1/e)$. Note that this angle may also be obtained as $\alpha = \arctan(b/a) = \arctan \sqrt{e^2 - 1}$.

A quantity analogous to the eccentric anomaly E must now be defined in order to derive Kepler's equation for hyperbolic orbits. In order to do this, we need to introduce the concept of hyperbolic sines and cosines. In the case of elliptic orbits, the equation of the orbit in Cartesian coordinates is given by

$$\frac{x^2}{a^2} + \frac{y^2}{b^2} = 1 \tag{7.24}$$

Using the geometry shown in Figure 7.1, we can write a parametric form of equation (7.24) in terms of E:

$$x = a \cos E \qquad y = b \sin E \tag{7.25}$$

The expressions for $\sin E$ and $\cos E$ can be rewritten in terms of exponential functions:

$$\cos E = \tfrac{1}{2}(e^{iE} + e^{-iE})$$

and

$$\sin E = \frac{e^{iE} - e^{-iE}}{2i} \tag{7.26}$$

It can be shown that the complex functions defined in equations (7.26) obey the exact relationships that define the trigonometric functions.

The hyperbolic sines and cosines (called sinh and cosh) can be defined by analogy using the equation for a hyperbola in Cartesian coordinates:

$$\frac{x^2}{a^2} - \frac{y^2}{b^2} = 1 \tag{7.27}$$

with

$$x = a \cosh F \qquad y = b \sinh F \tag{7.28}$$

where F is the hyperbolic eccentric anomaly. The hyperbolic sines and cosines can also be expressed in terms of exponential functions:

$$\cosh F = \tfrac{1}{2}(e^F + e^{-F})$$

and

$$\sinh F = \tfrac{1}{2}(e^F - e^{-F}) \tag{7.29}$$

The relation between the hyperbolic eccentric anomaly F and the true anomaly f [corresponding to equation (7.9)] is given by

$$r = \frac{a(e^2 - 1)}{1 + e \cos f} = a(e \cosh F - 1) \tag{7.30}$$

which may be obtained when the projections of the point on the hyperbola on the x and y axes are evaluated using

$$r \cos f = a(e - a \cosh F)$$

and

$$r \sin f = a\sqrt{e^2 - 1} \, \sinh F$$

Another form of the relation between f and F is

$$\tan(\tfrac{1}{2}f) = \left(\frac{e + 1}{e - 1}\right)^{1/2} \tanh(\tfrac{1}{2}F) \tag{7.31}$$

showing similarity to the corresponding relation applicable to elliptic orbits.

As expected, the eccentric anomaly for hyperbolas is related to an area, just as in the case of elliptic motion, it is an angle or arc. Consider the areas formed by the lines OP and OS and by the curve representing the hyperbola between P and S. Then the eccentric anomaly is given by

$$F = 2\frac{\text{Area}(POS)}{a^2}$$

Kepler's equation can now be derived for hyperbolic orbits in a manner similar to that used for the elliptic orbits. We start with equations (7.10) and (7.11), and then, using (7.30) in the same manner as for the ellipse and integrating part way along the path of the hyperbola, we obtain

$$(t_2 - t_1) = \frac{T}{2\pi}[e(\sinh F_2 - \sinh F_1) - (F_2 - F_1)] \qquad (7.32)$$

The parameter T is given by $2\pi \sqrt{a^3/GM}$, which is still valid even though a body moving in a hyperbolic orbit has no "period" in the usual sense.

A particularly interesting case is the parabolic orbit for which the eccentricity is precisely equal to 1. The polar equation of the parabola is

$$r = \frac{p}{1 + \cos f} = \frac{p}{1 + \cos \theta} \qquad (7.33)$$

where we have assumed that $\theta_0 = 0$. Once again, the conservation of angular momentum permits the derivation of an equation analogous to Kepler's equation. The point P in Figure 6.2 is called the "pericenter" of the parabola. At that point, the angular momentum is given as

$$L = mv_p r_p = mr^2\frac{d\theta}{dt} \qquad (7.34)$$

where v_p is the velocity of the mass m at the pericenter. From Figure 6.2, we have $r_p = \frac{1}{2}p$. The velocity at the pericenter, v_p, can be calculated using the fact that the total energy of the mass m in the orbit is always zero so that the kinetic energy is always precisely equal to the potential energy. At the pericenter

$$\tfrac{1}{2}mv_p^2 = \frac{GmM}{r_p} = \frac{2GmM}{p} \qquad (7.35)$$

Therefore,

$$v_p = 2\sqrt{\frac{GM}{p}}$$

and so the angular momentum is

$$L = m\sqrt{GMp} = mr^2\frac{d\theta}{dt} \tag{7.36}$$

and rewriting this equation so that it can be integrated part way around the orbit gives

$$\sqrt{GMP}\int_{t_0}^{t} dt = p^2\int_{0}^{\theta}\frac{d\theta}{(1+\cos\theta)^2} \tag{7.37}$$

integrating equation (7.37) yields

$$\sqrt{GMP}\,(t-t_0) = p^2\int_{0}^{\theta}\frac{d\theta}{(1+\cos\theta)^2} \tag{7.38}$$

where $t = t_0$ and $\theta = 0$ correspond to pericenter passage. The integral on the right side may be evaluated considering that $(1 + \cos\theta)^2 = 4\cos^4(\frac{1}{2}\theta)$ and that

$$\int\frac{dx}{\cos^4 x} = \int\frac{\sin^2 x}{\cos^4 x}\,dx + \int\frac{\cos^2 x}{\cos^4 x}\,dx$$

from which

$$\int\frac{dx}{\cos^4 x} = \tfrac{1}{3}\tan^3 x + \tan x$$

Consequently, the elapsed time on parabolic orbits becomes

$$2\sqrt{\frac{GM}{p^3}}\,(t-t_0) = \tan\left(\tfrac{1}{2}\theta\right) + \tfrac{1}{3}\tan^3\left(\tfrac{1}{2}\theta\right) \tag{7.39}$$

Note the similarity between the expression for the mean anomaly for elliptic orbits $l = 2\pi(t - t_0)$ and equation (7.39). For elliptic orbits, $2\pi/T = \sqrt{GM/a^3}$, and now the semilatus rectum takes over the role of the semimajor axis. Since the solution of cubic equations is known in explicit form, the time dependence of the true anomaly, and consequently the functional dependence of the radial distance on time, can be given in closed form. Equation (7.39) is known as Barker's equation and was published by Euler in 1743.

Equation (7.39) shows that, as $\theta \to \pi$, $t \to \infty$. Furthermore, the $p = 2rp$ relation for an ellipse is satisfied when $a(1 - e^2) = 2a(1 - e)$ or when $1 - e^2 = 0$, which gives $e = 1$ as expected. The same limit process applies to hyperbolic orbits.

Note that the term $\tan(\frac{1}{2}\theta)$ can also be used to express the radial distance, and equation (7.33) may be written as

$$r = \tfrac{1}{2}p[1 + \tan^2(\tfrac{1}{2}\theta)] \tag{7.40}$$

This derivation is concluded with a remark concerning the limit $e \to 1$ for conic sections. Parabolic orbits, besides the $e = 1$ condition, must also satisfy the $a \to \infty$ requirement. If the length of the semimajor axis is finite and $e = 1$, we have flat (straight-line) elliptic and hyperbolic orbits as mentioned before, since the length of the semiminor axis and the length of the semilatus rectum become zero.

Another note of considerable interest is that parabolic orbits display a singular property. By this we mean that all those orbits for which the total energy is negative ($E_T < 0$) are ellipses, and all those orbits for which the energy is positive ($E_T > 0$) are hyperbolas, but only one special value of the energy ($E_T = 0$) results in parabolic orbits. This fact has some interesting practical and theoretical consequences. Since the initial conditions are usually not known exactly in practical problems, it seldom happens that the energy constant is exactly zero. The same applies to the case when the nature of the orbit is established by observations, which furnish only approximations for the energy. In other words, instead of having exactly zero, we might have small positive or small negative values for the energy of the actual two-body orbit. The situation becomes even more complicated when the values of the physical constants that enter the energy equation are considered. Since the values of the constants of gravity and of the central mass are known only approximately (within error limits), the computed total energy will also be an approximation. We conclude that for practically important cases the constant of energy is determined only approximately. When this value is close to zero, the actual orbit might be an ellipse, a parabola, or a hyperbola. In order to find the nature of the orbit, several more observations are required, but even then the orbit often remains undetermined.

Besides these practical aspects, the theoretical implications might be mentioned briefly. We speak about instability when slight changes in the initial conditions result in greatly different orbits. This is the case of orbits for which the constant of energy is close to zero, since a small increase in speed will result in an escaping (hyperbolic) orbit and a small speed reduction gives an elliptic orbit. Such a case might lead us to the problem of nonpredictability of orbits. The situation is significantly complicated if some other forces enter the system besides those considered in the problem of two bodies, such as drag (which will slow down the orbit), propulsion (which might change the energy to positive or to negative values), perturbations due to other bodies, and so on.

For these theoretical reasons, parabolic orbits are of special interest, because they represent examples of nonpredictable orbits. Their practical significance is limited since they exist for only a highly special value of the energy.

These remarks concerning the examples of the limits of predictability should be compared to the notes appearing in the section on physical constants in Appendix 2.

The basic ideas are discussed by Gauss (1809) and Herget (1948). Many examples and additional geometric presentations are given in McCuskey's (1963) and Thomson's (1961) books. The special case of straight-line motion ($e = 1$, $b = 0$) is discussed systematically and in detail by Roy (1978).

Kepler's equation (7.20) is written in terms of the variable E, the eccentric anomaly, which is an angle measured from the geometric center of the ellipse shown in Figure 7.1. Sometimes, there are better variables that can be used for the practical determination of orbits, and the technique of developing the transformations from one of these variables to another is called "regularization." The objective of these transformations is often to eliminate the singularities from the equations of motion. The appearance of singularities is usually due to the fact that the gravitational force law introduces inverse square terms in the equations of motion. When the distances become small, these terms dominate and the accuracy of the computation might become questionable.

The credit for introducing regularizing variables usually is given to Sundman (1912), who introduced regularization in order to show the existence of solutions of the differential equations of motion. It is interesting to note that such pure mathematical exercises led to everyday practical techniques used today in our applied orbit mechanics. The combination of a mathematical existence proof and increased accuracy of numerical integration of the orbits of space probes represents an important message to promote the cooperation of engineers and mathematicians.

In orbit mechanics, of course, the distances never become zero, since collision (or impact) occurs before $r = 0$, due to the finite size of the bodies involved; nevertheless, the accuracy of the computation is reduced at close approaches, even when the numerical integration process allows the use of smaller time steps. The introduction of properly selected new variables regularizes the equations of motion, and accuracy can be maintained at the price of using transformations. Both the true and eccentric anomalies are such regularizing variables, and the transformation that introduce them are expressed as

$$df = \frac{L}{m} \frac{dt}{r^2} \qquad (7.41)$$

and

$$dE = a \frac{2\pi}{T} \frac{dt}{r} \qquad (7.42)$$

The first relation, as we have seen, is identical to the conservation of the angular momentum (7.34), and the second follows from Kepler's equation using (7.21).

Considering the above equations, which introduce the true and eccentric anomalies as new independent variables, the origin of regularization could be contributed to Kepler. He certainly had no idea of regularization, especially since, to him, calculus and differential equations were not known. Furthermore, Kepler's interest in the dynamics of the solar system excluded close approaches. Nevertheless, the use of the eccentric anomaly introduced in his equation allows us to increase the accuracy of our numerical integrations. We might consider this a demonstration of a true genius, or once again, we might celebrate the power of combination of two fields, that of Kepler's work to describe planetary motions and our efforts to compute accurate Earth-to-Moon trajectories. Ever since Kepler's work, the subject of regularization has been popular. Euler used it to study the straight-line motions in the problem of two bodies. Levi-Civita regularized the restricted problem of three bodies in 1903. Stumpff (1949) and Herrick (1965) used basically the same idea when they introduced the concept of universal variables.

The above transformations of the independent variable may be generalized to

$$ds = A \frac{dt}{r^m} \qquad (7.43)$$

or

$$ds = \frac{dt}{f(r)} \qquad (7.44)$$

where A and m are constants.

A recently introduced anomaly is known as the intermediate anomaly, which is given by $m = \frac{3}{2}$. Note that $m = 1$ gives the eccentric anomaly and $m = 2$ the true anomaly, with proper selection of the constant A. In general, the above transformations of the time introduce new independent variables that are often denoted by s. When the integration of the differential equations of motion is performed using s as the independent variable, the method is often called *s-integration*.

The above transformations, from a numerical point of view, might be considered analytic step regulations since as the distance r decreases and close approaches or collisions occur, the time step is to be reduced. Considering the above transformation equations, this means that the integration step using t will be reduced but the step size using s might be kept constant. The transformation regulates automatically and analytically the step size.

When transformations of the independent variable are combined with the proper transformations of the dependent variables and the conservation principles (energy and momentum) are used, the result is not only a regularized but also a linearized system of equations describing the gravitational problem of two bodies. As we have seen in Chapter 4, when the true anomaly is introduced as the new independent variable and $1/r = u$ is used as the new dependent variable, the differential equation of motion became a second-order linear differential equation representing a harmonic oscillator [equations (4.25) and (4.64)]. Because of the practical numerical importance of such "smoothing transformations," there are a large number of contributions in the literature of modern celestial mechanics dedicated to this subject.

The use of regularization in space dynamics becomes mandatory at close approaches occurring at departures, arrivals, and gravity-assist orbit maneuvers (which will be discussed in Chapter 8) and, in general, when orbit computations with high accuracy are required. In celestial mechanics, the accurate computation of cometary orbits also calls for regularization.

A modern comprehensive treatment of regularization and of the associated linearization is given in the book by Stiefel and Scheifele (1971).

7.2 DERIVATION OF LAMBERT'S THEOREM

In 1761, J. H. Lambert developed another formula that can be used to establish the parameters or orbits from certain measurements made from Earth. The original motivation was to determine the orbits of comets, and it is now known as *Lambert's theorem*. More recently, Lambert's theorem has been used to calculate the orbits of various spacecraft, including the Voyager interplanetary missions, which required very accurate determination of orbital parameters for the spacecraft to execute fly-bys of the outer planets. Remember that Neptune is almost 40 times as far away from the Sun as Earth. The Apollo spacecraft were also guided to their landing sites on the Moon using the Lambert Guidance Program.

Lambert's theorem can be derived from Kepler's equation (7.21), and it is generically the same kind of relationship. In the case of Kepler's equation, a time interval between two positions of a body in an orbit is measured, as well as the value of the eccentric anomaly at these two times. The eccentric anomaly is an angle measured from the geometric center of the orbit. The use of this coordinate system is sometimes inconvenient, and the virtue of Lambert's theorem is that the measurements that must be made to determine the orbital parameters can be performed from the focus of the elliptic orbit, which is more convenient for certain applications. Figure 7.3 shows the geometric situation. The body in orbit moves from point P_1 to point P_2 in the time interval $t_2 - t_1$. Lambert's theorem states that the time interval depends on two parameters, the sum $r_1 + r_2$ and the chord length C, as defined in Figure 7.3. In turn, $r_1 + r_2$ and C can be related to the eccentricity e and the semimajor axis a of the ellipse. We will start the derivation of Lambert's theorem from Kepler's equation. At the points P_1 and P_2, Kepler's equation takes the form

$$t_1 = \frac{T}{2\pi}(E_1 - e \sin E_1) \tag{7.45}$$

and

$$t_2 = \frac{T}{2\pi}(E_2 - e \sin E_2) \tag{7.46}$$

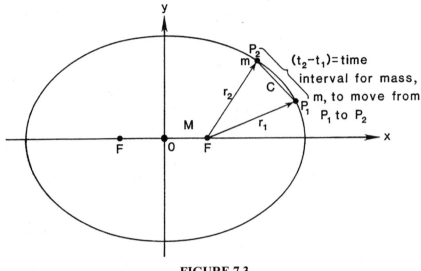

FIGURE 7.3

The time difference $t_2 - t_1$ is one of the measured quantities. Equations (7.45) and (7.46) represent the dynamics of the motion.

Turning to the geometry of the orbit shown in Figure 7.3, we can write the following equations for r_1, r_2, and C:

$$r_1 = a(1 - e \cos E_1) \qquad (7.47)$$

$$r_2 = a(1 - e \cos E_2) \qquad (7.48)$$

$$C^2 = a^2(\cos E_2 - \cos E_1)^2 + a^2(1 - e^2)(\sin E_2 - \sin E_1)^2 \qquad (7.49)$$

Equation (7.49) can be derived by referring to Figure 7.1. In terms of the x and y axes of the coordinate system, we can write

$$x = ae + FD = a \cos E \qquad y = QD \sin E = b \sin E \qquad (7.50)$$

because from Figure 7.1 it can be seen that $QD = BM$. In terms of the coordinate system at the focus of the ellipse,

$$x = FD = a \cos E - ae = a(\cos E - e) \qquad (7.51)$$

The chord length we want can be derived by looking at Figure 7.3:

$$C^2 = (x_1 - x_2)^2 + (y_1 - y_2)^2 \tag{7.52}$$

Using (7.50) and (7.5), we have, from (7.52),

$$C^2 = a^2(\cos E_2 - \cos E_1)^2 + b^2(\sin E_2 - \sin E_1)^2 \tag{7.53}$$

and, since $b = a\sqrt{1 - e^2}$, we can write

$$C^2 = a^2(\cos E_2 - \cos E_1)^2 + a^2(1 - e^2)(\sin E_2 - \sin E_1)^2 \tag{7.54}$$

which is identical to (7.49).

Having done this, we can now proceed with the remainder of the derivation by defining r_1 and r_2 in terms of the angles E_1 and E_2. Using equations (7.47) and (7.48), we have

$$r_1 + r_2 = a(1 - e \cos E_1) + a(1 - e \cos E_2)$$

or

$$r_1 + r_2 = 2a[1 - \tfrac{1}{2}e(\cos E_2 + \cos E_1)] \tag{7.55}$$

Using the familiar half-angle formulas for trigonometric functions, we can write

$$\cos E_2 + \cos E_1 = 2 \cos [\tfrac{1}{2}(E_2 + E_1)] \cos [\tfrac{1}{2}(E_2 - E_1)] \tag{7.56}$$

so that

$$r_1 + r_2 = 2a \{1 - e \cos [\tfrac{1}{2}(E_2 + E_1)] \cos [\tfrac{1}{2}(E_2 - E_1)]\} \tag{7.57}$$

This equation along with the expression for the chord length C, equation (7.54), results from an analysis of the geometry of the elliptic orbit.

We will now return to dynamic (that is, time-dependent) relationships for the remainder of the derivation. From the Kepler equations at P_1 (7.45) and P_2 (7.46), we can write

$$(t_2 - t_1) = \frac{T}{2\pi}[(E_2 - E_1) - e(\sin E_2 - \sin E_1)] \tag{7.58}$$

and using the half-angle formula

$$(\sin E_2 - \sin E_1) = 2 \cos \left[\tfrac{1}{2}(E_2 + E_1 0\} \sin \left[\tfrac{1}{2}(E_2 - E_1)\right] \quad (7.59)$$

we can rewrite Kepler's equation as follows:

$$(t_2 - t_1) = \frac{T}{2\pi}\{(E_2 - E_1) - 2e \sin \left[\tfrac{1}{2}(E_2 + E_1)\right] \cos \left[\tfrac{1}{2}(E_2 - E_1)\right]\} \quad (7.60)$$

Equation (7.60) represents the measured quantity from the dynamics of the system, that is, the moving asteroid, comet, or spacecraft. The measured quantities from the geometry of the ellipse are equation (7.40) for the sum of r_1 and r_2 and equation (7.54) for the chord length C. The latter can also be rewritten using the half-angle formulas

$$C^2 = 4a^2\sin^2 \left[\tfrac{1}{2}(E_2 + E_1)\right] \{1 - e^2\cos^2[\tfrac{1}{2}(E_2 - E_1)]\} \quad (7.61)$$

Equations (7.61), (7.60), and (7.57) can now be solved to yield Lambert's theorem. What we will want to do is to eliminate the eccentric anomalies E_1 and E_2 because these are constructs. The objective is to write the measured time difference $t_2 - t_1$ in terms of the measured quantities r_1, r_2, and C.

To simplify the expressions, we make the following change of variables:

$$x = e \cos \left[\tfrac{1}{2}(E_2 + E_1)\right] \qquad y = \tfrac{1}{2}(E_2 - E_1) \quad (7.62)$$

Using these, we can rewrite equations (7.60), (7.57), and (7.61) as follows:

$$t_2 - t_1 = \frac{T}{2\pi}(2y - 2x \sin y) \quad (7.63)$$

$$r_1 + r_2 = 2a(1 - x \cos y) \quad (7.64)$$

$$C^2 = 4a^2\sin^2 y(1 - x^2) \quad (7.65)$$

We now need to eliminate the quantities x and y from the above equations. From (7.64) and (7.65), we can write

$$r_1 + r_2 + C = 2a(1 - x \cos y + \sqrt{1 - x^2} \sin y) \tag{7.66}$$

$$r_1 + r_2 - C = 2a(1 - x \cos y - \sqrt{1 - x^2} \sin y) \tag{7.67}$$

A new variable change permits a solution

$$x = \cos[\tfrac{1}{2}(\alpha + \beta)] \qquad y = \tfrac{1}{2}(\alpha - \beta) \tag{7.68}$$

so that we can rewrite equations (7.66) and (7.67) as follows:

$$r_1 + r_2 + C = 2a \{1 - \cos[\tfrac{1}{2}(\alpha + \beta)] \cos[\tfrac{1}{2}(\alpha - \beta)]$$
$$+ \sin[\tfrac{1}{2}(\alpha + \beta)] \sin[\tfrac{1}{2}(\alpha - \beta)]\} \tag{7.69}$$

$$r_1 + r_2 - C = 2a \{1 - \cos[\tfrac{1}{2}(\alpha + \beta)] \cos[\tfrac{1}{2}(\alpha - \beta)]$$
$$- \sin[\tfrac{1}{2}(\alpha + \beta)] \sin[\tfrac{1}{2}(\alpha - \beta)]\} \tag{7.70}$$

We now apply the following half-angle formulas:

$$\cos[\tfrac{1}{2}(\alpha + \beta)] \cos[\tfrac{1}{2}(\alpha - \beta)] = \tfrac{1}{2}\cos\alpha + \tfrac{1}{2}\cos\beta \tag{7.71}$$

and

$$\sin[\tfrac{1}{2}(\alpha + \beta)] \sin[\tfrac{1}{2}(\alpha - \beta)] = \tfrac{1}{2}\cos\beta + \tfrac{1}{2}\cos\alpha \tag{7.72}$$

Using these formulas in equations (7.69) and (7.70), we obtain

$$r_1 + r_2 + C = 2a(1 - \cos\alpha) = 4a \sin^2(\tfrac{1}{2}\alpha) \tag{7.73}$$

and

$$r_1 + r_2 - C = 2a(1 - \cos\beta) = 4a \sin^2(\tfrac{1}{2}\beta) \tag{7.74}$$

Rewriting equation (7.63) in terms of the variables α and β defined in equations (7.68), we have

$$t_2 - t_1 = \frac{T}{2\pi} \{(\alpha - \beta) - 2 \cos[\tfrac{1}{2}(\alpha + \beta)] \sin[\tfrac{1}{2}(\alpha - \beta)]\} \tag{7.75}$$

and applying the half-angle formula,

$$2 \cos[\tfrac{1}{2}(\alpha + \beta)] \sin[\tfrac{1}{2}(\alpha - \beta)] = \sin \alpha - \sin \beta \qquad (7.76)$$

so that we can write

$$t_2 - t_1 = \frac{T}{2\pi}[(\alpha - \beta) - (\sin \alpha - \sin \beta)] \qquad (7.77)$$

where in equation (7.77) the parameters α and β are defined as follows from equations (7.73) and (7.74):

$$\sin \frac{\alpha}{2} = \frac{1}{2}\left(\frac{r_1 + r_2 + C}{a}\right)^{1/2} \qquad (7.78)$$

and

$$\sin \frac{\beta}{2} = \frac{1}{2}\left(\frac{r_1 + r_2 - C}{a}\right)^{1/2} \qquad (7.79)$$

From the relationships that have been derived here, it can be seen that the time interval $t_2 - t_1$ depends on a, $r_1 + r_2$, and C. Equation (7.77) is called Lambert's theorem along with the definitions of α and β in equations (7.78) and (7.79). Just as in the case for Kepler's equation, Lambert's theorem is a transcendental equation. Therefore, iterative techniques must be applied to solve for the parameters of the orbit in the general case.

A simple application of Lambert's theorem is when we assume that at P_1 in Figure 7.3 the true anomaly f_1 is zero and at P_2 it is $\tfrac{1}{2}\pi$. In that case

$$r_1 = a(1 - e) \qquad (7.80)$$

and

$$r_2 = a(1 - e^2) = p \qquad (7.81)$$

and the chord length C is given as

$$C = a(1 - e)\sqrt{2 + 2e + e^2} \qquad (7.82)$$

From these equations, we can evaluate α and β for substitution into Lambert's theorem (7.77). When this manipulation is performed, we have, for the time interval,

$$t_2 - t_1 = \frac{T}{2\pi}(\text{arc cos } e - e\sqrt{1 - e^2}) \qquad (7.83)$$

Thus we have two equations, (7.82) and (7.83), from which we can determine the orbital parameters. Once again, these must be solved numerically because (7.83) is a transcendental equation.

A similar formulation may be derived for hyperbolic orbits. The starting parameters are the sum of the radial distances, the length of the chord, and the semimajor axis of the hyperbola. The radial distances from equation (7.3) are

$$r_1 = a(e \cosh F_1 - 1) \qquad (7.84)$$

and

$$r_2 = a(e \cosh F_2 - 1) \qquad (7.85)$$

The terms $\sin(\alpha/2)$ and $\sin(\beta/2)$ used for the elliptic problem now become the corresponding hyperbolic functions:

$$\sinh \frac{\gamma}{2} = \frac{1}{2}\left(\frac{r_1 + r_2 + C}{a}\right)^{1/2} \qquad (7.86)$$

and

$$\sinh \frac{\delta}{2} = \frac{1}{2}\left(\frac{r_1 + r_2 - C}{a}\right)^{1/2} \qquad (7.87)$$

Kepler's equation for hyperbolic orbits [equation (7.5)] becomes

$$\frac{2\pi}{T}(t_2 - t_1) = e(\sinh F_2 - \sinh F_1) - (F_2 - F_1) \qquad (7.88)$$

and the length of the chord may now be written as

$$C = a[(\cosh F_2 - \cosh F_1)^2 + (e^2 - 1)(\sinh F_2 - \sinh F_1)^2]^{1/2} \qquad (7.89)$$

Lambert's theorem for hyperbolic orbits becomes

$$\frac{2\pi}{T}(t_2 - t_1) = \sinh \gamma - \sinh \delta - (\gamma - \delta) \qquad (7.90)$$

Lambert's theorem for parabolic orbits was given by Newton and by Euler. The derivation is straightforward, and it makes use of the fact that a parabola may be considered as a "stretched" ellipse for which the semimajor axis a becomes infinite. It is therefore possible to use small-angle approximations for α and β, which eliminates the transcendental terms in equation (7.77).

Equations (7.78) and (7.79) can now be written as

$$\sin \frac{\alpha}{2} = \frac{1}{2}\left(\frac{r_1 + r_2 + C}{a}\right)^{1/2} \approx \frac{\alpha}{2}$$

$$\alpha \approx \left(\frac{r_1 + r_2 + C}{a}\right)^{1/2} \qquad (7.91)$$

and

$$\sin \frac{\beta}{2} = \frac{1}{2}\left(\frac{r_1 + r_2 - C}{a}\right)^{1/2} \approx \frac{\beta}{2}$$

$$\beta \approx \left(\frac{r_1 + r_2 - C}{a}\right)^{1/2} \qquad (7.92)$$

Equations (7.91) and (7.92) can now be used to evaluate the expression for Lambert's theorem, (7.77). To do this, it is necessary to use the next term in the expansions of $\sin \alpha$ and $\sin \beta$; otherwise a null result is obtained:

$$\sin \alpha = \alpha - \tfrac{1}{6}\alpha^3 \qquad (7.93)$$

and

$$\sin \beta = \beta - \tfrac{1}{6}\beta^3 \qquad (7.94)$$

Substituting the expressions (7.93) and (7.94) in equation (7.77) yields

$$t_2 - t_1 = \frac{T}{2\pi}\left(\frac{\alpha^3}{6} - \frac{\beta^3}{6}\right) \qquad (7.95)$$

or, by rearranging terms and using expressions (7.91) and (7.92),

$$t_2 - t_1 = \frac{T}{12\pi}\left[\left(\frac{r_1 + r_2 + C}{a}\right)^{3/2} - \left(\frac{r_1 + r_2 - C}{a}\right)^{3/2}\right] \quad (7.96)$$

Finally, since a parabolic orbit has an "infinite" period T and an infinite semimajor axis a, we can use Kepler's third law to arrive at the final result, recognizing that the ratio

$$\frac{T}{2\pi}\frac{1}{a^{3/2}} = \frac{1}{\sqrt{GM}} \quad (7.97)$$

remains constant no matter how stretched the ellipse becomes. Thus, the time difference in equation (7.97) becomes

$$t_2 - t_1 = \frac{1}{6\sqrt{GM}}[(r_1 + r_2 + C)^{3/2} - (r_1 + r_2 - C)^{3/2}] \quad (7.98)$$

The sign of the second term becomes positive if the angle between r_1 and r_2 is larger than 180°, and it is negative, as given in equation (7.91), when $f_2 - f_1 < 180°$.

In conclusion, we note that Lambert's theorem can also be used to determine the radius of curvature of the orbit and in this way to find whether a planet is superior or inferior, that is, whether Earth or the planet is nearer to the Sun.

One of the outstanding, detailed treatments of Lambert's theorem is given in Plummer's book (1918). For applications to orbit determination techniques, see Bate, Mueller, and White (1971), where Lambert's problem is referred to as Gauss' problem. (Note that the year Gauss was born, Lambert died—1777.) Several practically useful modifications of the original formulation are offered in this reference, such as rendezvous, intercept, and so on. Special attention is directed to the astrodynamics applications of the *f* and *g* series (see Chapter 3, Example 3.1), in connection with the *Lambert–Gauss problem*. Important and recent applications can be found in Battin's 1964 book (Chapters 3 and 5) and in his 1987 book (Chapter 6), mentioned in Appendix III.

EXAMPLES

7.1. Compute the semimajor axis of Mars's orbit in astronomical units and kilometers using 1.9 years for its orbital period.

From Kepler's third law, we have

$$\frac{T_1^2}{T_2^2} = \frac{a_1^3}{a_2^3}$$

which, when applied to Earth and Mars, results in

$$a_M = a_E \left(\frac{T_M}{T_E} \right)^{2/3}$$

where $a_E = 1$, $T_E = 1$ year, and $T_M = 1.9$ years. This gives $a_M = 1.9^{2/3}$ = 1.534 a.u. = 229.6 × 10^6 km.

7.2. If the semimajor axis of a planet from the Sun is 2870 × 10^6 km, what is its orbital period in years? First we find the semimajor axis in astronomical units so that, with the help of the table of physical constants given in Appendix 2, we can identify the planet:

$$a = 2870 \times 10^6 \text{ km} = 1.496 \times 10^8 \ 2870 \times 10^6 = 19.18 \text{ a.u.}$$

This number corresponds to the planet Uranus. Its orbital period becomes

$$T_U = (19.18)^{3/2} = 84 \text{ years}$$

7.3. Using the basic data for the *Explorer 7* satellite, compute the mass of Earth in kilograms. The relation between the period and the semimajor axis is

$$T = 2\pi \left(\frac{a^3}{GM_E} \right)^{1/2}$$

where the symbol a in our case stands for the semimajor axis of the satellite $a = 7190.86$ km and $T = 1.684$ h. From the above equation, we have

$$M_E = \frac{4\pi^2 a^3}{GT^2} = 5.986 \times 10^{24} \text{ kg}$$

This result shows that the determination of planetary masses is not a simple matter, and it is strongly influenced by the observational accuracy.

7.4. Show that the average value of the distance between the focus and the ellipse is the semimajor axis, provided the averaging is performed with respect to the eccentric anomaly. Averaging with respect to the eccentric anomaly, we have

$$r_{ave} = \frac{1}{2\pi} \int_0^{2\pi} r \, dE$$

The relation between r and E is

$$r = a(1 - e \cos E)$$

After substitution and integration, we have

$$r_{ave} = a$$

If we average with respect to the true anomaly, we have

$$r_{ave} = \frac{1}{2\pi} \int_0^{2\pi} r \, df = \frac{p}{2\pi} \int_0^{2\pi} \frac{df}{1 + e \cos f}$$

This integral becomes

$$\int \frac{df}{1 + e \cos f} = \frac{2}{\sqrt{1 - e^2}} \arctan\left(\frac{1-e}{1+e}\right)^{1/2} \tan\frac{f}{2}$$

or using

$$\tan\frac{f}{2} = \left(\frac{1-e}{1+e}\right)^{1/2} \tan\frac{E}{2}$$

we have

$$r_{ave} = \frac{p}{2\pi} \frac{2}{\sqrt{1 - e^2}} \left[\frac{E}{2}\right]_0^{2\pi} = \frac{p}{\sqrt{1 - e^2}} = b$$

It is of interest to find that the time average, or the average with respect to the mean anomaly of r, is neither a nor b but

$$r_{ave} = \frac{1}{T} \int_0^T r \, dt = a(1 + \tfrac{1}{2}e^2)$$

This result may be obtained by differentiating Kepler's equation,

$$dt = \frac{dt}{dE}dE = (1 - e \cos E)\, dE\left(\frac{T}{2\pi}\right)$$

and using $r = a(1 - e \cos E)$ under the integral sign.

As a conclusion, we observe that the mean distance between the focus and elliptic orbit, in general, is not the length of the semimajor axis. For detailed discussion, see R. A. Serafin's article in *Celestial Mechanics,* Vol. 21, p. 351, 1980, and Taff's (1985) book (see Appendix).

7.5. The following example shows how the orbital elements of the hyperbolic orbit are established and how time computations are performed.

Consider the perigee velocity $v_p = 12$ km/s at an altitude $h = 1000$ km. First the orbit must be classified by using the energy equation

$$\tfrac{1}{2}mv^2 = \frac{GMm}{r} + E_T$$

After substituting $v = v_p$ and $r = r_p = h + R_E$, we obtain, for the constant of energy k, defined as $2E_T/m$ or GM/a,

$$k = v_p^2 - \frac{2GM}{h + R_E} = 35.95 \text{ km}^2/\text{s}^2$$

Note that the total energy of the probe per unit mass is $\tfrac{1}{2}k$ since the energy is

$$E_T = \tfrac{1}{2}mv_p^2 - \frac{GMm}{h + R_E}$$

Since energy is positive, the orbit is a hyperbola. The perigee distance is

$$h + R_E = a(e - 1)$$

and the semimajor axis is

$$a = \frac{GM}{k}$$

From these equations, we have $a = 11,087.62$ km and $e = 1.66544$. The semiminor axis and the semilatus rectum are

$$b = a\sqrt{e^2 - 1} \quad \text{and} \quad p = a(e^2 - 1)$$

or $b = 14,755.48$ km and $p = 19,665.98$ km.

The angle of the asymptotes is $\alpha = \arctan\sqrt{e^2 - 1} = 53°5'5''$. Note that, by means of the energy equation, the velocity may be computed at any point of the hyperbola:

$$v = \sqrt{GM}\left(\frac{2}{r} + \frac{1}{a}\right)^{1/2}$$

For instance, the velocity at infinity is obtained by the limit process $r \to \infty$. We have

$$v_\infty = \left(\frac{GM}{a}\right)^{1/2} = 5.995 \text{ km/s}$$

The circular velocity at the given altitude is

$$v_c = \left(\frac{GM}{r_p}\right)^{1/2} = 7.350 \text{ km/s}$$

which is, of course, smaller than the perigee velocity.

The escape velocity at the altitude of $h = 1000$ km is

$$v_e = \left(\frac{2GM}{r_p}\right)^{1/2} = 10.39 \text{ km/s}$$

which is smaller than the given perigee velocity. This is another way to establish the fact that the orbit is a hyperbola.

The velocity at the point where the semilatus rectum intersects the hyperbola ($r = p$ and $f = \frac{1}{2}\pi$) is

$$v_Q = \left[GM\left(\frac{2}{p} + \frac{1}{a}\right)\right]^{1/2} = 8.746 \text{ km/s}$$

The computed velocities in this case might be ordered as

$$v_p > v_e > v_Q > v_c > v_\infty$$

but note that, depending on the value of the eccentricity, these inequalities might change their order.

To evaluate the time of travel on a hyperbolic orbit, equation (7.32) is used. For instance, from perigee P to point Q, we have

$$t = \frac{T}{2\pi}[e(\sinh F) - F]$$

where T is obtained from the relation, $T = 2\pi\sqrt{a^3/GM}$, and the values of the hyperbolic eccentric anomaly F are computed from equation (7.31). At point Q, the true anomaly is $90°$ or $f = \frac{1}{2}\pi$ and F is to be computed from

$$\tan(\tfrac{1}{4}\pi) = \left(\frac{e+1}{e-1}\right)^{1/2} \tan(\tfrac{1}{2}F)$$

or

$$F = \log\left(\frac{\sqrt{e+1} + \sqrt{e-1}}{\sqrt{e+1} - \sqrt{e-1}}\right) = \log(e + \sqrt{e^2 - 1}) = 1.0976918$$

The value of F may also be obtained using equation (7.30). The left side of this equation for $f = \frac{1}{2}\pi$ gives $a(e^2 - 1)$, which is the length of the semilatus rectum. From equation (7.30), we have

$$e^2 - 1 = e\cosh F - 1 \quad \text{or} \quad \cosh F = e$$

and consequently, $F = \log(e + \sqrt{e^2 - 1})$, which is the same result as obtained above. The *hyperbolic mean motion* becomes $n = 1.94677$ rad/h, and equation (7.32) gives $t = 0.575487$ h $= 34$ min 31.75 s.

Note that when equation (7.32) is used, the term $\sinh F$ becomes, in our case, $\sqrt{e^2 - 1}$ since $\cosh^2 F - \sinh^2 F = 1$ and $\cosh F = e$. A quick approximate computation of the travel time uses the average velocity as the perigee and at point Q:

$$v_{ave} = \tfrac{1}{2}(v_Q + v_p) = 10.37 \text{ km/s}$$

The distance between points P and Q is

$$PQ = [(FP)^2 + (FQ)^2]^{1/2}$$

which after substitutions for $FP = a(e - 1)$ and for $FQ = a(e^2 - 1)$ becomes $PQ = r_p\sqrt{1 + (e + 1)^2} = 21{,}004.5$ km. The time to travel this distance with the above average velocity is $t' = PQ/v_{ave} = 0.5625$ h. Note that the error $t - t'$ amounts to 2.3%, which supports the idea of approximation, especially considering the fact that no hyperbolic functions were used to obtain the approximate result.

Another question we might ask in connection with our example is related to the design of lunar trajectories. Our probe moving in a hyperbolic orbit will be influenced mostly by Earth's gravitational field unless it approaches another celestial body such as the Moon. If the departure time is properly selected, the probe will approach the Moon, and the Moon's gravitational field will control part of the trajectory. In orbit mechanics, we refer to this formulation as the restricted problem of three bodies: Earth, the Moon, and the probe. The "restriction" comes from the fact that the probe, because of its small mass, does not influence the motion of Earth and of the Moon, but Earth and the Moon control the motion of the third body, which is the probe. In Chapter 13, this problem will be discussed in detail, but at this time, we will neglect the Moon's gravitational influence on the probe, and we will establish the time it takes to reach the distance of the Moon using our hyperbolic orbit. This transfer time will be compared to the time it takes to reach the lunar orbit on an elliptic Hohmann transfer orbit.

For the Earth–Moon distance, we will use the semimajor axis of the lunar orbit around Earth, $R_{EM} = 384{,}000$ km, which is only an approximation since the eccentricity of the lunar orbit is 0.0549.

Computation of the hyperbolic transfer time requires the values of a, e, and R_{EM}. Using the same hyperbolic orbit as before, we have to compute the value of F from $R_{EM} = a(e \cosh F - 1)$ and then the transfer time from the hyperbolic Kepler equation. In this way, we obtain

$$\cosh F = \frac{R_{EM} + a}{ae} = 21.41735$$

and $F = 3.7568$. Using the same value as before for $2\pi/T$ we obtain, for the transfer time, 16.3725 h. Note that the actual transfer time will be shorter since the Moon's gravitational effect will assist the trajectory. (The above computed time corresponds to the time reaching the lunar distance.)

It is of considerable interest to compare this time to an elliptic transfer time using a Hohmann orbit with a circular parking orbit at altitude $h = 1000$ km as before. The departure velocity will be determined so that the transfer orbit will be an ellipse with perigee altitude $R_E + h$ and apogee height R_{EM}. The semimajor axis of the transfer ellipse is

$$a = \tfrac{1}{2}(R_{EM} + R_E + h) = 195{,}889 \text{ km}$$

and the transfer time becomes

$$T = \pi \left(\frac{a^3}{GM} \right)^{1/2} = 119.84 \text{ h}$$

which is approximately 5 days. (Note that the hyperbolic transfer time was about 0.7 day.)

The arrival velocity at the Moon (once again neglecting the lunar gravitational effect) is

$$v_A = \left(2GM \frac{R_E + h}{R_{EM}} \right)^{1/2} \frac{1}{\sqrt{R_{EM} + R_E + h}} = 0.1976 \text{ km/s}$$

and the elliptic perigee velocity is

$$v_p = \left(2GM \frac{R_{EM}}{R_E + h} \right)^{1/2} \frac{1}{\sqrt{R_{EM} + R_E + h}} = 10.296 \text{ km/s}$$

The hyperbolic departure velocity is 12 km/s, and the arrival velocity is $v_A = 6.1664$ km/s, which is obtained from the equation of energy, using $r = R_{EM}$. The circular velocity at $h = 1000$ km is 7.3501 km/s, and the required velocity boost for the elliptic orbit at the perigee is $v_p - v_c = 2.946$ km/s. The circular velocity of the Moon is

$$v_{cM} = \left(\frac{GM}{R_{EM}} \right)^{1/2} = 1.018 \text{ km/s}$$

therefore, the vehicle arriving on an elliptic orbit would require a velocity boost of $v_{cM} - v_A = 0.8207$ km/s to keep up with the Moon. The probe arriving on a hyperbolic orbit has a higher velocity than the lunar velocity. The velocity vector of the probe arriving on the ellip-

EXAMPLE   **131**

tic transfer orbit is tangential to the Moon's orbit, but this is not the case for the hyperbolic transfer orbit.

Some details of the intersection of the hyperbolic transfer orbit and the Moon's orbit are shown in Figure 7.4. The two components of the arrival velocity of the probe are the radial dr/dt and the normal $r(df/dt)$ components. The normal component of the arrival velocity may be obtained from the momentum conservation, $L/m = r^2(df/dt)$ as $r(df/dt) = L/mr$, where $L/m = v_p r_p$ and $r = R_{EM}$. In this way, we have $r(df/dt) = 0.23033$ km/s. The direction of the arrival velocity (v_A) is given by the angle between v_A and dr/dt:

$$\lambda = \arcsin r\frac{df/dt}{v_a} = 2.1406°$$

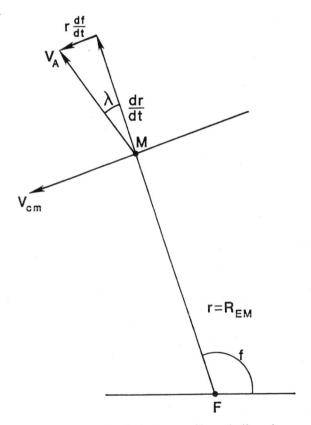

FIGURE 7.4 Arrival at the Moon on a hyperbolic trajectory.

The angle between the Moon's velocity v_{cM} (which is normal to r) and the probe's velocity, which is tangential to the hyperbola, is $90°$ $- \lambda$.

The hyperbolic orbit will intersect the lunar orbit when $F =$ 3.7568 at point M, as shown before. From this, the true anomaly becomes $f = 124.731°$, which may be obtained from equation (7.31) or (7.3) by computing f from

$$R_{EM} = \frac{a(e^2 - 1)}{1 + e \cos f}$$

The angle between the asymptote and the radial direction is $180 - (\alpha + f) = 2.1706°$.

7.6. The perigee velocities of elliptic and hyperbolic orbits having the same semimajor axes are related by

$$\frac{(v_e)_P}{(v_h)_P} = \left(\frac{e_h - 1}{1 - e_e}\right)^{1/2} \left(\frac{1 + e_e}{1 + e_h}\right)^{1/2}$$

where $(v_e)_P$ and $(v_h)_P$ are the perigee velocities of elliptic and hyperbolic orbits and e_e and e_h are the corresponding eccentricities. To show the above relation, consider the principle of energy conservation, which gives, for elliptic orbits,

$$v_e = \sqrt{GM}\left(\frac{2}{r} - \frac{1}{a}\right)^{1/2}$$

and for hyperbolic orbits

$$v_h = \sqrt{GM}\left(\frac{2}{r} + \frac{1}{a}\right)^{1/2}$$

The perigee distances are

$$r_e = a_e(1 - e_e) \text{ and } r_h = a_h(e_h - 1)$$

The perigee velocities become

$$(v_e)_P = \left(\frac{GM}{a_e}\right)^{1/2} \left(\frac{1 + e_e}{1 - e_e}\right)^{1/2}$$

and

$$(v_h)_P = \left(\frac{GM}{a_h} \right)^{1/2} \left(\frac{e_e + 1}{e_e - 1} \right)^{1/2}$$

If $a_h = a_e$, the required result is obtained from the last two equations. If the eccentricities are related by

$$e_e = 1 - x \qquad e_h = 1 + x$$

where $0 \leq x \leq 1$, then the inequality $v_e < v_h$ is satisfied. The physical meaning of this is that if the eccentricity of the elliptic orbit is less than 1 by the quantity x, and the eccentricity of the hyperbolic orbit is larger than 1 by the same amount, then the hyperbolic perigee velocity is larger than the corresponding elliptic perigee velocity, as expected. For instance, if $x = 0.5$ and $e_e = 0.5$, $e_h = 1.5$, we have, for $a_h = a_e$, that

$$\frac{(v_e)_P}{(v_h)_P} = \sqrt{0.6} \quad \text{and} \quad (v_e)_P < (v_h)_P$$

The proof of the above result is obtained by substituting $e_e = 1 - x$ and $e_h = 1 + x$ into the equation given for $(v_c)_P/(v_h)_P$. In this way, we obtain

$$\frac{(v_e)_P}{(v_h)_P} = \left(\frac{2 - x}{2 + x} \right)^{1/2}$$

which quantity is always less than 1.

Note that if the eccentricities are close to 1, that is, x is much less than 1, the velocity ratio becomes approximately $1 - \frac{1}{2}x$.

7.7. In case the reader is interested in additional numerical examples, the following results are offered without details.

If the circular parking orbit has a period of 2 h, then the altitude is 1681 km, and the circular velocity is 7.033 km/s. The escape velocity at this altitude is 9.946 km/s. To obtain a hyperbolic transfer orbit to the Moon, let us use a velocity that is 10% higher than the escape velocity (i.e., 10.941 km/s).

To establish the transfer hyperbola, we now use the above altitude of 1681 km and the velocity of 10.941 km/s, which is tangential to the parking orbit at the time of departure. The semimajor axis of the hyperbolic orbit is 19,177.6 km, its eccentricity $e = 1.42$, its semiminor axis is 19,340.4 km, and the velocity at infinite distance becomes 4.56 km/s. The transfer time is 20 h, 34 min, 29 s. The angle of the asymptotes $\alpha = 45.233°$.

7.8. The reader who likes to manipulate equations will enjoy the following exercise.

For a probe near Earth, we have a tangential velocity v_P (normal to the radius vector from the center of Earth) at an altitude h_P. First we establish the ranges of the v_P values that result in elliptic, parabolic, and hyperbolic orbits. Using $r_P = h_P + R_E$ for the distance from the center of Earth to the probe, the energy conservation gives

$$\tfrac{1}{2}mv_P^2 - \frac{GMm}{r_P} = E_T$$

where m is the mass of the probe and M is the mass of Earth. Elliptic orbits correspond to $E_T < 0$, parabolic orbits to $E_T = 0$, and hyperbolic orbits to $E_T > 0$. Consequently, for elliptic orbits,

$$0 \le v_P < \left(2\frac{GM}{r_P}\right)^{1/2}$$

for parabolic orbits,

$$v_P = \left(2\frac{GM}{r_P}\right)^{1/2}$$

and for hyperbolic orbits,

$$\left(2\frac{GM}{r_P}\right)^{1/2} < v_P$$

As a second part of this exercise, we establish the semimajor axes, eccentricities, and semilatus rectums (*recta,* if the reader is a Latinist) as functions of v_P, GM, and r_P for the elliptic, parabolic, and hyperbolic orbits. For the elliptic orbits,

$$a = \frac{GMr_P}{2GM - r_P v_P^2} \qquad e = \frac{r_P v_P^2}{GM} - 1$$

and

$$p = \frac{r^2_P v^2_P}{GM}.$$

The first result follows from the energy conservation equation:

$$v^2_P = GM\left(\frac{2}{r_P} - \frac{1}{a}\right)$$

the second from using $r_P = a(1 - e)$ and the third from $p = a(1 - e^2)$. For parabolic orbits, $a = \infty$, $e = 1$, $p = 2r_P$. For hyperbolic orbits,

$$a = \frac{MGr_P}{r_P v^2_P - 2MG}$$

$$e = \frac{r^2_P v^2_P}{GM} - 1$$

and

$$p = \frac{r^2_P v^2_P}{GM}.$$

As the third part of this exercise, we find the velocities for the three cases at point Q (see Figures 4.1 and 7.1). Note that point Q is located at the intersection of the semilatus rectum with the orbit.

The velocity at Q is obtained from the energy conservation equation, which for elliptic orbits becomes

$$v^2_Q = GM\left(\frac{2}{p} - \frac{1}{a}\right)$$

The result for elliptic orbits is

$$v^2_Q = \frac{2}{r_P}\left(\frac{GM}{r_P v^2_P} - 1\right) + v^2_P$$

which is the same for hyperbolic orbits. For parabolic orbits, we have

$$v^2_Q = \frac{G}{r_P}$$

The last part of this exercise is the computation of the elapsed times between perigee passage and arrival at point Q. For elliptic orbits, Kepler's equation gives

$$\frac{2\pi}{T}t = E - e \sin E$$

where the mean motion is given by $2\pi/T = \sqrt{GM/a^3}$. The eccentric anomaly E at perigee is zero and at point Q is obtained from the value of the true anomaly ($f = \frac{1}{2}\pi$) using equation (7.9). In this way, we have, at point Q,

$$E = \arcsin \sqrt{1 - e^2}$$

and the transfer time becomes

$$T = \left(\frac{a^3}{GM}\right)^{1/2} (\arcsin\sqrt{1 - e^2} - e\sqrt{1 - e^2})$$

In this result, a and e are to be expressed as functions of GM, r_P, and v_P as obtained in the second part of this exercise. For parabolic orbits, the required result is given by equation (7.8):

$$\left(\frac{GM}{p^3}\right)^{1/2} t = \tan\frac{f}{2} + \frac{1}{3}\tan^3\frac{f}{2}$$

Once again, at perigee $f = 0$ and $t = 0$, and at point Q, $f = \frac{1}{2}\pi$; therefore,

$$T = \frac{8}{3}\left(\frac{2r_P^3}{GM}\right)^{1/2}$$

The case of hyperbolic orbits is similar to the one discussed in connection with elliptic orbits, but now Kepler's equation becomes

$$\frac{2\pi}{T}t = e(\sinh F) - F$$

where the mean motion is still $\sqrt{GM/a^3}$, and the hyperbolic eccentric anomaly F is related to the true anomaly by equation (7.3):

$$\frac{p}{1 + e \cos f} = a(e \cosh F - 1)$$

At the perigee $f = 0$, and from equation (7.3), we have $\cosh F = 1$, or $F = 0$. At point Q, $f = \frac{1}{2}\pi$ and $\cosh F = e$, or $\sinh F = \sqrt{e^2 - 1}$. Using the inverse hyperbolic cosine function, we have

$$T = \left(\frac{a^3}{GM}\right)^{1/2}[e\sqrt{e^2 - 1} - \log(e + \sqrt{e^2 - 1})]$$

In this result, a and e are still to be expressed as functions of r_P, v_P, and GM. These relations were found in the second part of this exercise.

7.9. As a simple exercise, the previously computed time of travel on a hyperbolic orbit shown in Figure 7.2 between points P and Q may be computed using Lambert's formulation. The radial distances in the case of a hyperbolic orbit with perigee altitude $h = 1000$ km and velocity $v_P = 12$ km/s are

$$r_1 = r_P = h + R_E = 7378.14 \text{ km}$$

and

$$r_2 = p = a(e^2 - 1) = 19{,}666 \text{ km}$$

The previously obtained values for the semimajor axis and eccentricity were $a = 11{,}087.62$ km and $e = 1.66544$. The length of the chord is

$$C = \sqrt{r_1^2 + r_2^2} = 21{,}004.5 \text{ km}$$

and $\sinh(\frac{1}{2}\gamma) = 1.0409$, $\sinh(\frac{1}{2}\gamma) = 0.3690$. The corresponding values for γ and δ may be obtained using the previously given logarithmic formulas: $\gamma = 1.8199$ and $\delta = 0.7223$. Substitution in equation (7.2), using $T = 3.226$ h, as before, gives $t_2 - t_1 = 0.575$ h.

PROBLEMS

7.1. Compute the average value of the orbital velocity for elliptic motion. Perform the averaging with respect to the eccentric anomaly, true anomaly, and mean anomaly.

7.2. Show that for small values of eccentricity, when e^3 and higher powers can be neglected, the solution of Kepler's equation becomes

$$E = l + e(1 + e \cos l)\sin l$$

7.3. Find the values of the eccentric and true anomalies when $l = \frac{1}{4}\pi, \frac{1}{2}\pi, \frac{3}{4}\pi, \pi$ for an elliptic orbit with eccentricity $e = 0.2$.

7.4. Show that $df/dt = l/mr^2$, $dE/dt = 2ma/Tr$, and $dl/dt = 2\pi/T$ on elliptic orbits.

7.5. Derive formulas for the radial (dr/dt) and normal $[r(d\theta/dt)]$ velocity components as function of GM, the semimajor axis, eccentricity, and true anomaly for elliptic and hyperbolic orbits.

7.6. Derive the formula for the velocity

$$v = \sqrt{\left(\frac{dr}{dt}\right)^2 + \left(r\frac{d\theta}{dt}\right)^2}$$

also as a function of GM, a, e, and ϕ, and compare the result obtained from the equation of energy, once again for elliptic and hyperbolic orbits.

7.7. Derive formulas for dr/df, dr/dE, and dr/dl for elliptic and hyperbolic motions and compute their values at the pericenter.

7.8. A space probe is launched vertically up, from the surface of Earth, with speed u. Neglecting drag, compute its height, when its velocity becomes zero, as a function of u. When the probe reaches this height, it is given a transverse velocity v. Find the nature and the parameters of the orbit and show how these depend on u and v.

CHAPTER 8

ORBITAL MANEUVERING
OF SPACECRAFT

Since spaceflight began with the launching of *Sputnik I* by the Soviets on October 4, 1957, thousands of spacecraft have been launched. Complex maneuvers such as the docking of two spacecraft have been performed both in Earth orbit and in lunar orbit. Spacecraft in Earth orbit have been moved from one orbit to another in order to achieve certain objectives. Finally, spacecraft have been sent to all of the planets of the solar system except Pluto. All of these operations require a detailed understanding of how spacecraft behave when orbital changes are made. We will see that the important quantity is the velocity change needed to make the desired change. This is normally called Δv (delta vee), and it is the purpose of this chapter to show how Δv calculations are performed for various orbital changes.

The first example we shall consider is the calculation of the velocity change required to change the orbital plane of an Earth-orbiting satellite. The situation is illustrated in Figure 8.1. Let the angle between the planes be θ, and let us assume that both orbits are circular with radius R.

From Figure 8.1, it is clear that the velocity vector addition that needs to be performed to change the orbital plane is shown in Figure 8.2. Since the orbital velocity depends only on the radius of the circular orbit,

$$v = \sqrt{\frac{GM_E}{R}} \tag{8.1}$$

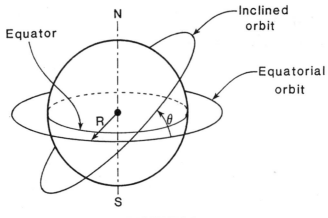

FIGURE 8.1

where M_E is the mass of Earth, the magnitude of the velocity does not change.

The direction change in the vector can be calculated from Figure 8.1:

$$\tfrac{1}{2}|\Delta\mathbf{v}| = |\mathbf{v}|\sin(\tfrac{1}{2}\theta) \tag{8.2}$$

To illustrate what is required, let us look at the case when $\theta = \tfrac{1}{2}\pi$, that is, when we want to change an equatorial to a polar orbit. In that case, equation (8.2) becomes

$$|\Delta\mathbf{v}| = 2|\mathbf{v}|\frac{1}{\sqrt{2}} = \sqrt{2}|\mathbf{v}| \tag{8.3}$$

Thus, the velocity increment necessary to execute this change is larger than the orbital velocity. Since the energy required is proportional to the

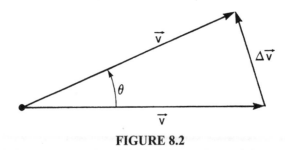

FIGURE 8.2

square of the velocity, the energy necessary to change a circular equatorial orbit to a polar orbit is twice that required to place the satellite into Earth orbit (with radius R) in the first place.

Another useful approximation to consider is the case when the angle θ is small. In that case,

$$\sin(\tfrac{1}{2}\theta) \approx \tfrac{1}{2}\theta \tag{8.4}$$

Thus, we have

$$|\Delta \mathbf{v}| \approx |\mathbf{v}|\theta \tag{8.5}$$

which says that the magnitude of the velocity increment is proportional to the angle θ measured in radians. Therefore, an orbital plane change of 10° (approximately 0.17 rad) requires a velocity change of 17% of the orbital velocity, or an energy change of about 34% of the orbital kinetic energy. From these considerations, it is clear that orbital plane changes are difficult to achieve since they require large velocity (and therefore energy) changes. This is a consequence of the fact that orbital velocities are large.

We will now turn to the case in which the satellite moves from one orbit to another in the same plane. A common maneuver is to "circularize" the orbit of an Earth-orbiting satellite that has been launched into an elliptic orbit. This situation is illustrated in Figure 8.3. The satellite is originally in an elliptic orbit with a semimajor axis a and an eccentricity e. Using equation (6.11) in Chapter 6, we have, for the total energy of the orbit,

$$E_T = -\frac{GM_E m}{2a} \tag{8.6}$$

Assume that Earth is located at the right-hand focus of the ellipse; we can therefore write, for the total energy at perigee, r_p,

$$E_T = -\frac{GM_E m}{2a} = \frac{1}{2}mv_p^2 - \frac{GM_E m}{r_p} \tag{8.7}$$

From Figure 8.3, we have

$$r_p = a(1 - e) \tag{8.8}$$

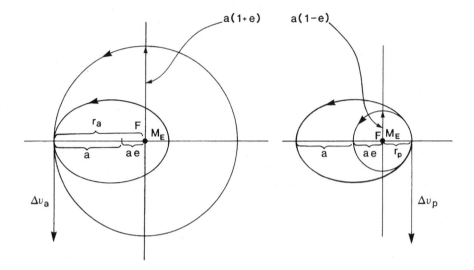

Circularization at apogee   Circularization at perigee

FIGURE 8.3

and using this relationship in equation (8.7), we can solve for the velocity:

$$v_p^2 = \frac{GM_E}{a}\left[\frac{1+e}{1-e}\right] \tag{8.9}$$

A similar calculation yields, for the velocity at apogee, v_a,

$$v_a^2 = \frac{GM_E}{a}\left[\frac{1-e}{1+e}\right] \tag{8.10}$$

Circularization of the orbit at perigee means that we change the velocity at that point so that the object (mass m) executes a circular orbit with a radius r_p. The velocity in the circular orbit would be

$$v_p^2(\text{circular}) = \frac{GM_E}{r_p} = \frac{GM_E}{a(1-e)} \tag{8.11}$$

from which it is clear that

$$v_p > v_p(\text{circular}) \tag{8.12}$$

Therefore, the spacecraft must be slowed down if the orbit is circularized at perigee.

Normally, orbit circularization for Earth-orbiting spacecraft is carried out at apogee, in which case the spacecraft must be speeded up. At apogee, the velocity in the elliptic orbit is

$$v_a^2 = \frac{GM_E(1-e)}{a(1+e)} \tag{8.13}$$

and the velocity in a circular orbit with radius $a(1+e)$ is

$$v_a^2(\text{circular}) = \frac{GM_E}{a(1+e)} \tag{8.14}$$

so that

$$v_a(\text{circular}) > v_a \tag{8.15}$$

The velocity increment that must be supplied to circularize the orbit at apogee is

$$\Delta v_a = v_a(\text{circular}) - v_a$$

$$= \sqrt{\frac{GM_E}{a}} \left[\frac{1}{\sqrt{1+e}} - \frac{\sqrt{1-e}}{\sqrt{1+e}} \right] \tag{8.16}$$

A useful approximation can be developed if the eccentricity e is small. In that case, we assume that $1/\sqrt{1+e}$ and $\sqrt{1-e}$ can both be replaced by $1 - \frac{1}{2}e$. Thus

$$\Delta v_a \cong \sqrt{\frac{GM_E}{a}} [(1 - \tfrac{1}{2}e) - (1 - \tfrac{1}{2}e)^2] \tag{8.17}$$

$$\Delta v_a \cong \sqrt{\frac{GM_E}{a}} \left(\frac{e}{2} \right) = v_a(\text{circular}) \frac{e}{2}$$

Thus, the velocity increment is proportional to the eccentricity of the original orbit into which the spacecraft was placed. The smaller the eccentricity, the smaller the velocity increment necessary for circularization.

Another important calculation is to determine the velocity increments that are necessary for escape from Earth orbit. These are necessary for spacecraft that are intended to fly to the Moon or to other planets, because for all practical purposes these must reach escape velocity to execute their missions. In this case, also, the procedure will be to add tangential velocity increments either at apogee or at perigee. We will now show that for all values of the eccentricity e of the orbit the velocity increment to escape from perigee is smaller than that necessary to escape from apogee. Thus, the most advantageous place from which to leave Earth is actually the point in the elliptic orbit that is closest to Earth.

To show this result, we will start by calculating the escape velocities at apogee and at perigee by using the fact that for an escape trajectory the total energy of the orbit is zero. Therefore, we have

$$\tfrac{1}{2} m v_a^2 (\text{escape}) = \frac{GM_E}{a(1+e)} \tag{8.18}$$

and

$$\tfrac{1}{2} m v_p^2 (\text{escape}) = \frac{GM_E}{a(1-e)} \tag{8.19}$$

Therefore, we have, for the velocity increments at apogee and at perigee,

$$\Delta v_a = v_a(\text{escape}) - v_a$$

$$\Delta v_a = \sqrt{\frac{GM_E}{a}} \left(\frac{\sqrt{2}}{\sqrt{1+e}} - \frac{\sqrt{1-e}}{\sqrt{1+e}} \right) \tag{8.20}$$

and

$$\Delta v_p = v_p(\text{escape}) - v_p$$

$$= \sqrt{\frac{GM_E}{a}} \left(\frac{\sqrt{2}}{\sqrt{1-e}} - \frac{\sqrt{1+e}}{\sqrt{1-e}} \right) \tag{8.21}$$

From these relationships, it follows that for any value of e between 0 and 1, that is, for any bound orbit,

$$\Delta v_p < \Delta v_a \tag{8.22}$$

Therefore, it is easier to escape from an elliptic orbit at perigee than at apogee. Quantitatively, it is once again useful to see how the escape velocities behave if the value of the eccentricity e of the orbit is small. In this case

$$\Delta v_a = \sqrt{\frac{GM_E}{a}}[\sqrt{2}(1 - \tfrac{1}{2}e) - (1 - \tfrac{1}{2}e)^2]$$

$$= \sqrt{\frac{GM_E}{a}}\left[\sqrt{2} - \frac{e}{\sqrt{2}} - 1 + e - \tfrac{1}{4}e^2\right] \tag{8.23}$$

Ignoring the term in e^2, we have

$$\Delta v_a = \sqrt{\frac{GM_E}{a}}\left[(\sqrt{2} - 1) + e\left(\frac{\sqrt{2} - 1}{\sqrt{2}}\right)\right]$$

$$= \sqrt{\frac{GM_E}{a}}(\sqrt{2} - 1)\left(1 + \frac{e}{\sqrt{2}}\right) \tag{8.24}$$

Likewise, for perigee we have

$$\Delta v_p = \sqrt{\frac{GM_E}{a}}[\sqrt{2}(1 + \tfrac{1}{2}e) - (1 + \tfrac{1}{2}e)^2]$$

$$= \sqrt{\frac{GM_E}{a}}\left[\sqrt{2} + \frac{e}{\sqrt{2}} - 1 - e - \tfrac{1}{4}e^2\right] \tag{8.25}$$

Again, dropping the term in e^2, we obtain

$$\Delta v_p = \sqrt{\frac{GM_E}{a}}\left[1 - \frac{e}{\sqrt{2}}\right](\sqrt{2} - 1) \tag{8.26}$$

From equations (8.26) and (8.24), we can write the following expression for the difference between the velocity increments at perigee and apogee

$$\Delta v_p - \Delta v_a = (v_{escape} - v_{orbital})\sqrt{2}e \tag{8.27}$$

where v_{escape} and $v_{orbital}$ are the escape and orbital velocities for a circular orbit with radius a. As would be expected, the greater the eccentricity of the orbit from which the escape is to be performed, the larger difference between the escape delta-vee at perigee (Δv_p) and apogee (Δv_a).

We will now look at the problem of making a transfer from one approximately circular orbit to another. This is a very common maneuver. In the case of Earth-orbiting spacecraft, an orbital transfer maneuver is most often executed when a spacecraft is placed in a geosynchronous orbit. In this case, the spacecraft starts from a near-circular orbit close to Earth's surface. [This is often called a low earth orbit (LEO).] The spacecraft, along with an appropriate propulsion stage, is placed in Earth orbit, usually about 200 miles above Earth's surface, by an appropriate booster rocket. Once this orbit is achieved, the propulsion stage, called the transfer stage, is fired, and the spacecraft is placed into an elliptic orbit designed to reach an altitude appropriate for the geosynchronous orbit. Once the spacecraft is at the correct altitude, the transfer stage is fired again to circularize the orbit at an altitude and inclination so that it is geosynchronous.

Orbital transfers of this kind are also used for spacecraft intended to fly to other planets in the solar system. In this case, the first step is also to place the spacecraft in a LEO with an appropriate booster rocket. The next step is to reach a transfer orbit that will take the spacecraft to the target planet. In this case, the transfer orbit must be an orbit around the Sun, rather than Earth, which puts much greater demands on the transfer stage. If the spacecraft is being sent to a planet that is farther away from the Sun than Earth, then the velocity of the spacecraft must be increased to reach the planet. The velocity of the spacecraft in this case must actually be larger than the velocity of Earth in its orbit around the Sun. If the target planet is closer to the Sun than to Earth, then the final velocity of the spacecraft will be smaller than the velocity of Earth in its orbit. The spacecraft thus "falls" toward the Sun, and it intersects the orbit of the target planet on its new trajectory.

The general situation is illustrated in Figure 8.4. What we want to do is to move the spacecraft from a circular orbit with radius R_1 to another circular orbit with a radius R_2. It turns out that the most efficient way, that is, the way that requires the least energy, is to add a velocity increment at point A that puts the satellite in an elliptic orbit that is tangential to the new circular orbit with radius R_2 at the point B. From Figure 8.4, it is clear that we need to put the spacecraft in an elliptic orbit with a semimajor axis equal to a to execute the maneuver from point A to point B. The velocity of the spacecraft in a circular orbit of radius R_1 moving around the mass M is

$$v^2(R_1) = \frac{GM}{R_1} \qquad (8.28)$$

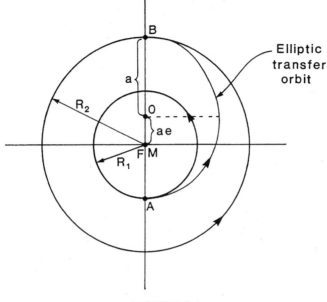

FIGURE 8.4

From equation (8.9), the velocity at perigee is

$$v_p^2 = \frac{GM}{a}\left(\frac{1+e}{1-e}\right) \tag{8.29}$$

where e is the eccentricity of the ellipse, a is the semimajor axis, and the mass M is located at the focus of the ellipse. From Figure 8.4, the following relationships can be derived:

$$R_2 = a + ae \quad R_1 = a - ae \tag{8.30}$$

Therefore, we have

$$R_1 + R_2 = 2a \tag{8.31}$$

and we can solve for the quantities $1 + e$ and $1 - e$ as follows:

$$1 + e = \frac{R_2}{a} \quad \text{and} \quad 1 - e = \frac{R_1}{a} \tag{8.32}$$

Substituting equations (8.31) and (8.32) for the velocity at perigee gives

$$v_p^2 = \frac{2GM}{R_1 + R_2}\left(\frac{R_2}{R_1}\right) \tag{8.33}$$

The velocity increment required to move the spacecraft from point A to point B is therefore

$$\Delta v(A \rightarrow B) = v_p - v(R_1)$$

which when the algebra is done becomes

$$\Delta v(A \rightarrow B) = \sqrt{\frac{GM}{R_1}}\sqrt{\frac{2R_2}{R_1 + R_2}} - \sqrt{\frac{GM}{R_1}}$$

$$= v(R_1)\left[\sqrt{\frac{2R_2}{R_1 + R_2}} - 1\right] \tag{8.34}$$

An interesting case to examine is to look at the velocity increment required for a spacecraft to travel from Earth to Jupiter using equation (8.34) and neglecting the mass of Jupiter. In this case, Earth is located at point A and Jupiter is at point B. Jupiter is 5.2 miles as far away from the Sun as Earth. The factor in the bracket on the right side of equation (8.34) is therefore

$$\left[\sqrt{\frac{2R_2}{R_1 + R_2}} - 1\right] = \sqrt{\frac{10.4}{6.2}} - 1 = 0.29 \tag{8.35}$$

The orbital velocity of Earth around the Sun, which is defined as $v(R_1)$ here, is about 107,229 km/h. Therefore, the velocity increment required for this maneuver is

$$\Delta v(\text{Earth} \rightarrow \text{Jupiter}) = 31,096 \text{ km/h}$$

We shall return to this point in more detail in the next chapter.

A second case that we should discuss is the placement of a spacecraft in geosynchronous orbit. Referring back to Chapter 2, we will take 6371 km as the radius of Earth and 42,639 km as the radius of the geosynchronous orbit as measured from the center of Earth. The velocity increment at point A can be calculated using equation (8.34). In this case, we have

$$R_1 = 6371 + 322 = 6693 \text{ km}$$
$$R_2 = 42{,}639 \text{ km} \qquad (8.36)$$
$$v(R_1) = 27{,}750 \text{ km}$$

where we have assumed that the starting LEO is at an altitude of 322 km. Thus,

$$\Delta v(A \rightarrow B) = 27{,}750 \left(\sqrt{\frac{85.2}{49.3}} - 1 \right)$$

$$= 27{,}750 \times 0.31 = 8{,}603 \text{ km/h} \qquad (8.37)$$

This, however, is not enough. The spacecraft has reached point B with this velocity increment, but now the orbit needs to be circularized at the radius R_2. The velocity increment required to circularize the orbit at R_2 (42,639 km) is given by equation (8.16). This can be rewritten as follows in the notation of Figure 8.4:

$$\Delta v_A(\text{circularize}) = \sqrt{\frac{GM}{R_1}} \left[\sqrt{\frac{R_1}{R_2}} \left(1 - \frac{2\sqrt{R_1}}{\sqrt{R_1 + R_2}} \right) \right]$$

$$= 27{,}750 \, [0.39(1 - 0.74)] = 2{,}813 \text{ km/h} \qquad (8.38)$$

Therefore, this velocity increment must be added to the one shown in equation (8.37) to get the entire velocity increment for the transfer:

$$\Delta v(\text{total}) = \Delta v(A \rightarrow B) + \Delta v_a(\text{circularize}) = 11{,}416 \text{ km/h} \qquad (8.39)$$

Note that this velocity increment is quite substantial, about 40% of that required to place a payload into Earth orbit. Finally, the general expression for the velocity change required to change from one circular orbit, R_1, to another one of radius R_2 is

$$\Delta v(\text{total}) = \sqrt{\frac{GM}{R_1}} \left[\left(\sqrt{\frac{2R_2}{R_1 + R_2}} - 1 \right) + \sqrt{\frac{R_1}{R_2}} \left(1 - \sqrt{\frac{2R_2}{R_1 + R_2}} \right) \right] \qquad (8.40)$$

The periods of the orbits involved in the transfers we have discussed can be determined using Kepler's third law of planetary motion. Referring back to Figure 8.4, the period of the transfer orbit is

$$\frac{R^3}{T^2} = \frac{GM}{4\pi^2} \qquad (8.41)$$

In the case of the transfer orbit, we equate R to the semimajor axis of the ellipse shown in Figure 8.4. To write this in terms of R_1 and R_2, we use equation (8.31), and substituting into (8.40), we have

$$\frac{(R_1 + R_2)^3}{2^3 T^2} = \frac{GM}{4\pi^2} \tag{8.42}$$

The period of the orbit can now be expressed as follows:

$$T = \sqrt{\frac{\pi^2 (R_1 + R_2)^3}{2GM}} \tag{8.43}$$

This equation can be evaluated for both the trip to Jupiter and the transfer to geosynchronous orbit. Since only half the orbit needs to be executed to make the transfer, the trip time to Jupiter is about 30 months and the time to reach geosynchronous orbit is about 12 h.

The transfer orbit described in Figure 8.4 was first suggested by W. Hohmann in 1925. It can be shown by a simple argument that the Hohmann transfer orbit, for most cases of practical interest, is the one that minimizes the energy necessary to make the transfer. This can easily be seen by referring to Figure 8.5. By definition, the Hohmann orbit is tangent to the circle with radius R_2 at point B. The total energy of the Hohmann orbit is given as

$$E_T(\text{Hohmann}) = -\frac{GM}{2a} \tag{8.44}$$

Let us now look at an orbit that is slightly displaced from the Hohmann orbit, which also reaches the orbit or radius R_2 but crosses it before the point B (see the dashed curve in Figure 8.5). This orbit can also transfer a spacecraft to the orbit with radius R_2, but its total energy will be different because the semimajor axis of the new orbit will be somewhat larger. We define the length of the line BB' as Δa. Since the point A is fixed, the center of the ellipse, point 0, also moves by approximately $\Delta a/2$ so that the total energy of the new orbit is approximately

$$E_T(\text{new orbit}) = -\frac{GM}{2a + \Delta a/2} \tag{8.45}$$

The total energy of this orbit is larger than that for the Hohmann transfer orbit because it is *less negative* when equation (8.45) is compared to (8.44). If the spacecraft does not leave the inside circle (with radius R_1),

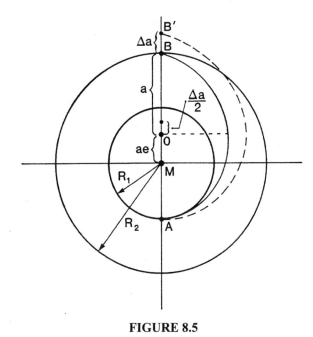

FIGURE 8.5

with a tangential orbit, then similar arguments can be made to show that the total energy is also larger. Obviously, for smaller total energies (that is, smaller values of a), the spacecraft will never reach the point B. Of course, these arguments are valid only if the gravitational attractions of other bodies in the neighborhood can be neglected.

W. Hohmann's original work was published in his book, *Die Erreich-barkeit der Himmelkörper,* Oldenburg Publ., Munich (1925) (English translation published by NASA, Washington, D.C., 1960). Note that the exact proof of Hohmann's result appeared only in 1961, showing that the optimum condition is limited by the requirement that $R_2/R_1 < 11.94$. See Ehricke's Vol. 2, Chapter 6 (1962) and Battin's Sections 9.2 and 11.3 (1987), mentioned in the Appendix. An easy-to-read report with engineering orientation is by J. B. Eades, "Orbital Transfer," *Bulletin of the Virginia Polytechnic Institute,* Blacksburg, VA, 1965.

EXAMPLES

8.1. What velocity increment would be needed to have Earth escape from the Sun? The average velocity of Earth around the Sun is given by

$$v_E = \left[\frac{G(m_s + m_E)}{a_E} \right]^{1/2} = 29.78 \text{ km/s}$$

The escape velocity is

$$v_c = \sqrt{2}\, v_E = 42.12 \text{ km/s}$$

The required velocity increment is

$$v_e - v_E = 12.34 \text{ km/s}$$

8.2. If the eccentricity of Earth's orbit is $e = 0.0167$, what is the ratio of the perihelion and aphelion velocities? Using equations (8.9) and (8.10), we have

$$\frac{v_P}{v_a} = \frac{1.0167}{0.9833} = 1.03397$$

Using the approximate formula,

$$\frac{v_P}{v_a} \approx 1 + 2e$$

we have 1.0334. The difference between the two velocities is obtained from equations (8.9) and (8.10):

$$v_P - v_a = \left(\frac{GM}{a} \right)^{1/2} \left[\left(\frac{1+e}{1-e} \right)^{1/2} - \left(\frac{1-e}{1+e} \right)^{1/2} \right]$$

or

$$\Delta v = \left(\frac{GM}{a} \right)^{1/2} \frac{2e}{\sqrt{(1-e^2)}}$$

The numerical value is $0.0334 \times 29.8 = 0.995$ km/s. Note that an approximate equation for the velocity difference valid for small values of the eccentricity is

$$\Delta v \approx 2e \left(\frac{GM}{a} \right)^{1/2}$$

which in our case gives the same result, that is, $\Delta v = 0.995$ km/s.

8.3. The escape velocity from a planet is computed from equation (5.31):

$$v_e = \left(\frac{2GM}{r}\right)^{1/2}$$

where r is the distance from the center of the planet to the escaping body. The escape velocities from the surfaces of planets become

Planet	Mercury	Venus	Earth	Mars	Jupiter	Saturn	Uranus	Neptune	Pluto
Escape velocity (km/s)	4.25	10.36	11.18	5.024	59.57	35.56	21.33	23.76	1.16

Note that planetary atmospheres will influence the above-computed escape velocities. The escape velocity from the surface of the Sun is 617.53 km/s and from the Moon is 2.375 km/s. The escape velocity from the surface of Earth is 11.18 km/s. A stone thrown at a velocity of 1.118 km/s will not escape unless we are on a planet that is smaller than Earth. The formula for the velocity of escape is

$$v_c = \left(\frac{2GM_P}{R_P}\right)^{1/2} = R_P(\tfrac{8}{3}\pi G\rho_p)^{1/2}$$

where ρ_p is the density of the planet. This new formula is obtained by writing $(4\pi/3)R_P^3\rho_p$ for the mass of the planet. Since the mean density of Earth is 5.52 g/cm³, the radius of a planet (with Earth's density) should be $R_P = R_E/10 = 637.8$ km to have 1.118 km/s as the escape velocity from the surface. Note that the escape velocity is proportional to the planetary radius as long as the density and the gravitational constant are not altered.

8.4. In this example, a Hohmann transfer orbit is established from a circular parking orbit to a geosynchronous orbit.

If the velocity is increased tangentially on the circular parking orbit, the point where this occurs will be the perigee of the transfer orbit. The circular velocity of a satellite at elevation $h = 300$ km is

$$v_c = \left(\frac{GM}{R_E + h}\right)^{1/2} = 7.726 \text{ km/s}$$

The elliptic orbit's apogee distance from Earth's center is $d = 42240.14$ km, and the semimajor axis of the transfer orbit is

$$a = \tfrac{1}{2}(h + R_E + d) = 24459.14 \text{ km}$$

The eccentricity of the transfer orbit is

$$e = \frac{d - R_E - h}{d + R_E + h} = 0.727$$

which follows from the expression for the perigee distance,

$$r_P = a(1 - e)$$

where $r_P = h + R_e$ and $a = \tfrac{1}{2}(h + R_e + d)$. The perigee and apogee velocities are

$$v_p = \left(\frac{GM}{a}\right)^{1/2}\left(\frac{1-e}{1+e}\right)^{1/2} = \left(\frac{2GMd}{(d + R_e + h)(R_e + h)}\right)^{1/2} = 10.153 \text{ km/s}$$

and

$$v_A = \left(\frac{GM}{a}\right)^{1/2}\left(\frac{1-e}{1+e}\right)^{1/2} = \left(\frac{2GM(R_E + h)}{d(d + R_E + h)}\right)^{1/2} = 1.605 \text{ km/s}$$

The radial distance of the original circular parking orbit from the center of Earth is $R_e + h = 6678.14$ km, and the original circular velocity is $v_c = 7.726$ km/s. The velocity increment at perigee is $\Delta v_P = v_P - v_c = 2.427$ km/s. The circular velocity at apogee distance is $v_c' = \sqrt{GM/d} = 3.072$ km/s, but the velocity on the elliptic orbit at apogee is smaller than this. Therefore, if a circular orbit is to be obtained at apogee, we need a velocity increment of $\Delta v_A = v_c - v_A = 1.467$ km/s. The total velocity change is $\Delta v_t = \Delta v_P + \Delta v_A = 3.894$ km/s. The transfer time is half of the period of the elliptic orbit:

$$\tfrac{1}{2}T = \pi\left(\frac{a^3}{GM}\right)^{1/2} = 5 \text{ h } 17 \text{ min } 14.6 \text{ s}$$

8.5. The idea of the elliptic Hohmann transfer orbit can be used to establish a lunar trajectory. Note that the following equations offer only approximations since the Moon's effect on the trajectory is neglected. The parking orbit is at elevation h, so the circular velocity is

$$v_c = \left(\frac{GM}{h + R_e} \right)^{1/2}$$

The semimajor axis of the transfer orbit is

$$a = \tfrac{1}{2}(R_e + h + d_{EM})$$

where d_{EM} is the distance between the centers of Earth and the Moon. The eccentricity of the transfer orbit is

$$e = \frac{d_{EM} - h - R_e}{d_{EM} + h + R_e}$$

The perigee velocity is

$$v_P = \left(\frac{2GMd_{EM}}{(h + R_E)(h + R_E + d_{EM})} \right)^{1/2}$$

and the velocity increment on the parking orbit becomes

$$\Delta v_P = v_P - v_c$$

The arrival velocity at the Moon (neglecting in the first approximation the gravitational effect of the Moon) is the apogee velocity of the elliptic transfer orbit,

$$v_A = v_P \frac{h + R_e}{d_{EM}}$$

which equation utilizes the conservation of the angular momentum. Since $h + R_e < d_{EM}$, $v_A < v_P$, as expected. The circular velocity at apogee is the lunar velocity, which is higher than the arrival velocity; therefore, a new boost is required:

$$\Delta v_A = v_M - v_A$$

where v_M is the Moon's (approximate) circular velocity. The transfer time is once again half the period of the transfer ellipse:

$$\tfrac{1}{2}T = \pi \left(\frac{(h + R_E + d_{EM})^3}{8GM} \right)^{1/2}$$

8.6. This example inverts the usual problem of orbit changes and intends to establish the new orbit when a certain velocity change occurs. If the velocity increases from circular velocity (v_c) to v, then the change is $\Delta v = v - v_c$, or $v = v_c(1 + \Delta v/v_c)$. The dimensionless velocity change is $x = \Delta v/v_c$, which, when multiplied by 100, represents the percentage change of the velocity. If the circular velocity is increased, it becomes the perigee velocity, or

$$v_P = v_c(1 + x)$$

The circular velocity at perigee distance was

$$v_c = \left(\frac{GM}{a(1-e)} \right)^{1/2}$$

and the increased perigee velocity is

$$v_P = \left(\frac{GM(1+e)}{a(1-e)} \right)^{1/2}$$

After substitution of the expressions for v_c and v_P into the relation $v_P = v_c(1 + x)$, we have

$$\sqrt{1+e} = 1 + x$$

or

$$e = x(2 + x) \approx 2x$$

where the approximation is valid for small x and small e values. A 10% velocity increase on the circular orbit results in an elliptic orbit with eccentricity $e = 0.1(2 + 0.1) = 0.21$. The percentage velocity increase required for escape can be obtained when the relation $\sqrt{1+e} = 1 + x$ is used with $e = 1$. The result is $x = \sqrt{2} - 1$, that is, 41.42% increase (or more) of the circular velocity is required for escape. The orbital parameters of the elliptic orbit (when $x < 0.4142$) can be expressed as functions of x. For instance, if the altitude of the original circular orbit is h, the semimajor axis of the elliptic orbit becomes

$$a = \frac{h + R_e}{1 - 2x - x^2}$$

For a 10% velocity increase, the semimajor axis becomes $a = 1.266(h + R_e)$. The apogee velocity becomes

$$v_A = \frac{1-e}{1+e} v_P$$

or

$$v_A = \frac{1 - x(2 + x)}{(1 + x)^2} v_P$$

or

$$v_A = \frac{1 - x(2 + x)}{1 + x} v_c$$

For a 10% velocity increase, the apogee velocity becomes $0.7182v_c$.

8.7. The orbit of the artificial satellite *Explorer 6*, known as 1959 δ2, has a perigee distance of $r_P = 6622.6$ km and an apogee distance of $r_A = 48201$ km. (Note that these are not heights but distances from the center of Earth to a point on the orbit.) The semimajor axis is

$$a = \tfrac{1}{2}(r_P + r_A) = 27411.8 \text{ km}$$

The eccentricity is computed by using the equations for the perigee and apogee distances:

$$r_P = a(1 - e) \quad \text{and} \quad r_A = a(1 + e)$$

From these we have

$$e = \frac{r_A - r_P}{r_A + r_P} = 0.758$$

The semiminor axis and the semilatus rectum are

$$b = a\sqrt{1 - e^2} = 17{,}879 \text{ km}$$

and

$$p = a(1 - e^2) = 11{,}662 \text{ km}$$

The mean motion n is computed from Kepler's law:

$$GM_E = \left| \frac{2r}{T} \right|^2 a^3$$

From this we have $n = 1.4 \times 10^{-4}$ rad/s $= 12.1$ rad/day. The orbital period of the satellite is

$$T = \frac{2\pi}{n} = 12.5 \text{ h}$$

The perigee and apogee velocities are

$$v_P = \left(\frac{GM}{a} \right)^{1/2} \left(\frac{1+e}{1-e} \right)^{1/2} = 10.28 \text{ km/s}$$

and

$$v_A = \left(\frac{GM}{a} \right)^{1/2} \left(\frac{1-e}{1+e} \right)^{1/2} = 1.42 \text{ km/s}$$

Note that once the perigee velocity is known, the apogee velocity can be computed by

$$v_A = v_P \frac{1-e}{1+e}$$

The advantage of using this method is the simplicity of the formula. The disadvantage is that if the computed value of v_P is in error, v_A also will be in error. If this satellite is at perigee when we start observing time (i.e., at $t = 0, f = l = E = 0$; see the table of anomalies given in Chapter 7), its location at $t = 62.5$ min as it moves counterclockwise can be established as follows. The mean anomaly at this time is $l = nt = 0.526$ rad $= 30°$. The eccentric anomaly can be computed from

$$l = E - e \sin E$$

by iteration, starting with $E_0 = 30°$. The solution is $E \approx 71.27°$. The true anomaly is computed from

$$\tan \frac{f}{2} = \left(\frac{1+e}{1-e}\right)^{1/2} \tan \frac{E}{2}$$

giving $f = 125.271°$. The distance from the center of Earth can be computed either from $r = a/(1 - e \cos E)$ or from $r = p/(1 + e \cos f)$. The results are 20739.75 and 20739.93 km, showing an error of 0.177 km or less than 0.001%. This error is due to the error of the approximate solution obtained by solving Kepler's transcendental equation by iteration. In order to circularize the orbit at apogee, a velocity boost is required. This circular velocity is

$$(v_A)_c = \left(\frac{GM}{r_A}\right)^{1/2} = \left(\frac{GM}{a}\right)^{1/2} \frac{1}{\sqrt{1+e}} = 2.876 \text{ km/s}$$

and the velocity increment is

$$(v_A)_c = 2.876 - 1.42 = 1.456 \text{ km/s}$$

To circularize at perigee, the satellite will have to slow down. The circular velocity at perigee altitude is

$$(v_P)_c = \left(\frac{GM}{r_P}\right)^{1/2} = \left(\frac{GM}{a}\right)^{1/2} \frac{1}{\sqrt{1-e}} = 7.75 \text{ km/s}$$

and the velocity change is

$$(v_P)_e = 10.28 - 7.75 = 2.53 \text{ km/s}$$

As expected, the required velocity change is larger at perigee than at apogee; therefore, circularization is to be performed at apogee. To escape at apogee, the velocity must be

$$(v_A)_e = \sqrt{2}(v_A)_c = 4.07 \text{ km/s}$$

The increase of velocity is

$$(\Delta v_A)_e = 4.07 - 1.42 = 2.65 \text{ km/s}$$

The escape velocity at perigee is

$$(v_P)_e = 2(v_p)_c = 10.96 \text{ km/s}$$

The increase of velocity is

$$(v_P)_e = 10.96 - 10.28 = 0.68 \text{ km/s}$$

Note that $(\Delta v_A)_e > (\Delta v_P)_e$, and therefore it is more efficient to execute the escape maneuver at perigee than at apogee.

PROBLEMS

8.1. A space vehicle approaching a planet on a hyperbolic orbit (relative to the planet) wishes to be captured. Find the change (reduction) of the (hyperbolic) velocity at the pericenter in order to obtain an elliptic orbit with eccentricity e. The hyperbolic excess velocity of the probe is v_∞ and the distance of the pericenter is r_P.

8.2. A space probe is orbiting a planet on an elliptic orbit with apocenter r_A and pericenter r_P. Find the change of the pericenter if the velocity at the apocenter is increased by Δv_A.

CHAPTER 9

ELEMENTS OF
SPACECRAFT DYNAMICS

We have examined how spacecraft can be maneuvered from one orbit to another in the previous chapter. Before we can actually navigate a spacecraft, we must also understand how to orient the spacecraft with respect to a fixed point in space and how to control the orientation so that the vehicle can be stabilized with respect to a defined coordinate system. Control of spacecraft orientation is vital to assure that the propulsion system points in the right direction when it is activated to move the spacecraft from one orbit to another. Every time the space shuttle is maneuvered to initiate the procedures to return from orbit, the attitude of the vehicle must be accurately fixed so that when the orbital maneuvering engines are fired, it assumes the proper return trajectory. In addition, stabilization is essential if spacecraft are to carry out the functions for which they are designed. Accurate pointing is required for spacecraft such as *Landsat,* which is designed to take high-resolution pictures of the ground. The Hubble Space Telescope requires accurate pointing to do astronomy and the Voyager spacecraft also had to have accurate pointing capability in order to take high-resolution pictures of the outer planets.

   In order to understand how spacecraft can be controlled, we will develop some of the basic relationships governing the dynamics of rigid bodies. This is not meant to be a comprehensive treatment of the subject, but it will be enough so that the spacecraft maneuvers we will deal with in the remaining chapters can be understood. Essentially, what we will do is to develop transformations between a coordinate system that is fixed in the spacecraft (the rigid body in this case), which we will call the "s" system and an inertial coordinate system "i" fixed in space. To describe the motion of the spacecraft in inertial space (the i system) in response to the control system mounted on the spacecraft (the s system), we will have to derive the transformation equations between the two coordinate systems. Once these relationships are established, we will then apply them to understand the behavior of several simple spacecraft control systems.

We will start by analyzing the problem in two dimensions and then extend it to three. Figure 9.1 shows the i system and the s system with the appropriate unit vectors. The vector $\mathbf{r}$ is a position vector that defines a point in the spacecraft making an angle α with the unit vector $\hat{s}_1$. The components of the $\mathbf{r}$ vector in the s system are therefore

$$\mathbf{r} = \hat{s}_1 r \cos \alpha + \hat{s}_2 r \sin \alpha \qquad (9.1)$$

where r is the magnitude of the vector $\mathbf{r}$. The most common situation we must deal with is when the spacecraft is rotating in inertial space. Thus, we will assume that the s system is rotating with respect to the i system with a constant angular velocity $\boldsymbol{\omega}(s, i)$ defined by

$$\boldsymbol{\omega}(s, i) = \omega(s, i) \, \hat{s}_3 \qquad (9.2)$$

where $\hat{s}_3$ is the unit vector in the s system perpendicular to $\hat{s}_1$ and $\hat{s}_2$.

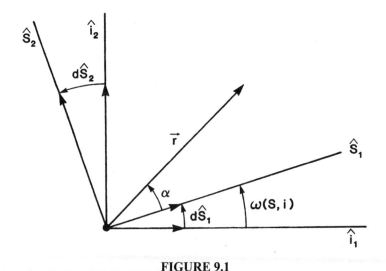

FIGURE 9.1

We now want to calculate the velocity vector in the s system and in the i system. For the s system we have

$$\frac{d\mathbf{r}}{dt}(s) = -\hat{s}_1 r \sin \alpha \frac{d\alpha}{dt} + \hat{s}_2 r \cos \alpha \frac{d\alpha}{dt} \tag{9.3}$$

where we have assumed that the magnitude of the vector $\mathbf{r}$ is constant and that we adopt the following convention:

$$\frac{d}{dt}(s) = \text{time derivative in } s \text{ system}$$

$$\tag{9.4}$$

$$\frac{d}{dt}(i) = \text{time derivative in } i \text{ system}$$

We will need to write the velocity vector in the inertial system i rather than the spacecraft system. In this case, we have to include how the unit vectors $\hat{s}_1$ and $\hat{s}_2$ behave as a function of time. Thus

$$\mathbf{v}(i) = \frac{d\mathbf{r}}{dt}(i) = \frac{d\mathbf{r}}{dt}(s) + \frac{d\hat{s}_1}{dt} r \cos \alpha + \frac{d\hat{s}_2}{dt} r \sin \alpha \tag{9.5}$$

The derivatives of the unit vectors can be evaluated by looking at Figure 9.1. By inspection, we can see that

$$d\hat{s}_1 = \hat{s}_2 \omega(s, i) \, dt \tag{9.6}$$

$$d\hat{s}_2 = \hat{s}_1 \omega(s, i) \, dt \tag{9.7}$$

The velocity vector in the i system is therefore

$$\mathbf{v}(i) = \frac{d\mathbf{r}}{dt}(i) = \frac{d\mathbf{r}}{dt}(s) - \hat{s}_1 \omega(s, i) r \sin \alpha + \hat{s}_2 \omega(s, i) r \cos \alpha \tag{9.8}$$

We now introduce the vector relationships in three dimensions in order to rewrite equation (9.8) in the most general form. The vector products of any system of three unit vectors are defined as follows:

$$\hat{s}_1 \times \hat{s}_1 = \hat{s}_2 \times \hat{s}_2 = \hat{s}_3 \times \hat{s}_3 = 0 \tag{9.9}$$

and

$$\hat{s}_1 \times \hat{s}_2 = \hat{s}_3 \qquad \hat{s}_2 \times \hat{s}_3 = \hat{s}_1 \qquad \hat{s}_3 \times \hat{s}_1 = \hat{s}_2$$
$$\hat{s}_2 \times \hat{s}_1 = -\hat{s}_3 \qquad \hat{s}_3 \times \hat{s}_2 = -\hat{s}_1 \qquad \hat{s}_1 \times \hat{s}_3 = -\hat{s}_2 \qquad (9.10)$$

We can now apply the relationships (9.10) to rewrite equation (9.8) as follows:

$$\omega(s, i) \times \mathbf{r} = \omega(s, i)\hat{s}_3 \times (\hat{s}_1 r \cos \alpha + \hat{s}_2 r \sin \alpha)$$
$$\hat{s}_2 \omega(s, i) r \cos \alpha - \hat{s}_1 \omega(s, i) r \sin \alpha \qquad (9.11)$$

Therefore, the velocity vector in the inertial system is

$$\mathbf{v}(i) = \frac{d\mathbf{r}}{dt}(i) = \frac{d\mathbf{r}}{dt}(s) + \omega(s, i) \times \mathbf{r} \qquad (9.12)$$

Expression (9.12) is quite general and applies to any vector in the i system. When the acceleration is calculated by taking the time derivative of the velocity, the second term in (9.12) yields the centripetal and the Coriolis accelerations.

We are now ready to rewrite Newton's second law of motion so that it can be used to describe the motion of the spacecraft when external forces are applied. The second law of motion is a universal relation between the applied force and the momentum:

$$\mathbf{F} = \frac{d\mathbf{P}}{dt} = m\frac{d\mathbf{v}}{dt} \qquad (9.13)$$

where $\mathbf{P}$ is the linear momentum and m is the mass of the object to which the force is applied. In the case of the point masses that we considered in Chapter 2, the angular momentum was defined as

$$\mathbf{L} = \mathbf{r} \times \mathbf{P} \qquad (9.14)$$

There is a relationship between the angular momentum and the vector product of the vector $\mathbf{r}$ and the force $\mathbf{F}$, called the torque, which has the same form as equation (9.13):

$$\mathbf{T} = \mathbf{r} \times \mathbf{F} = \mathbf{r} \times \frac{d\mathbf{P}}{dt} = \mathbf{r} \times m\frac{d\mathbf{v}}{dt} \qquad (9.15)$$

In Chapter 2, when we were discussing circular orbits of point masses, the angular velocity ω of the mass around the force center was defined as

$$\omega = \frac{v}{r} \tag{9.16}$$

where all quantities are scalars. Actually, all of the quantities in equation (9.16) are vectors, but in the special case of the circular orbits we considered, the three vectors $\mathbf{r}$, $\mathbf{v}$, and $\boldsymbol{\omega}$ are mutually perpendicular. The general relationship between the linear velocity and the angular velocity is

$$\mathbf{v} = \boldsymbol{\omega} \times \mathbf{r} \tag{9.17}$$

Equation (9.17) can be used to rewrite the second law of motion (9.15) as follows:

$$\mathbf{T} = \mathbf{r} \times \mathbf{F} = \mathbf{r} \times \frac{d\mathbf{P}}{dt} = \frac{d(\mathbf{r} \times \mathbf{P})}{dt} = \frac{d\mathbf{L}}{dt} \tag{9.18}$$

which can be done because for a rigid body the vector $\mathbf{r}$ is constant. Thus, the relationship between the force and the linear momentum (9.13) has the same form as the relationship between the torque and the angular momentum shown in equation (9.18). The angular momentum is defined in equation (9.14) which, using equation (9.17), can be rewritten as

$$\mathbf{L} = \mathbf{r} \times \mathbf{P} = m(\mathbf{r} \times \mathbf{v}) = m\mathbf{r} \times (\boldsymbol{\omega} \times \mathbf{r}) \tag{9.19}$$

It is possible to simplify this relationship by defining the moment of inertia I, which permits us to write the angular momentum as a function of the angular velocity in analogy with the relationship between the linear velocity and the linear momentum:

$$\mathbf{P} = m\mathbf{v} \tag{9.20}$$

For the angular momentum and the angular velocity we have

$$\mathbf{L} = I\boldsymbol{\omega} \tag{9.21}$$

where I is the moment of inertia.

There is a very important difference between equation (9.20) and (9.21). In the case of equation (9.20), the linear momentum vector and the velocity vector **v** always point in the same direction because the mass is a scalar quantity. The same is not true in the case of the angular momentum and the angular velocity. The angular momentum and the angular velocity need not point in the same direction, as can be seen by looking at equation (9.19). In the special case of the point mass moving in a circular orbit that we have already mentioned, the vectors **r**, **v**, and **ω** are mutually perpendicular, and we have

$$L = mr^2\omega \qquad (9.22)$$

where, by using equation (9.21), the moment of inertia becomes

$$I = mr^2 \qquad (9.23)$$

This is the familiar moment of inertia of a point mass m moving on a moment arm r. For objects that are more complex than a point mass, there are special cases where the angular momentum and the angular velocity are always colinear. For example, **L** and **ω** always point in the same direction for spherically symmetric objects. For objects that are not spherically symmetric, **L** and **ω** are colinear if the object rotates around one of the "principal axes." It is beyond the scope of this work to develop a general theory of the motion of rigid bodies to describe what happens in the general case. Almost all spacecraft are designed to be axially symmetric so that a general theory is not required.

Mathematically, the situation just described means that the moment of inertia is not a scalar but a tensor. A tensor that defines the relationship between two three-component vectors as indicated in (9.21) has nine components. The rules of matrix multiplication apply, and using these, equation (9.21) can be rewritten as follows:

$$\begin{pmatrix} L_1 \\ L_2 \\ L_3 \end{pmatrix} = \begin{pmatrix} I_{11} & I_{12} & I_{13} \\ I_{21} & I_{22} & I_{23} \\ I_{31} & I_{32} & I_{33} \end{pmatrix} \begin{pmatrix} \omega_1 \\ \omega_2 \\ \omega_3 \end{pmatrix} \qquad (9.24)$$

It is obvious from (9.24) that the vectors **L** and **ω** need not point in the same direction when the matrix multiplication is performed. There is another point that is also important. Unlike the mass in equation (9.20), which is always a constant, the value of the components of the moment of inertia may change depending on the choice of coordinate system.

Both of the mathematical difficulties we have just described can be dealt with more easily because the s coordinate system located in the spacecraft moves along with the spacecraft, and this means that the components of the moment-of-inertia tensor are constants. A further simplification results if the axes of the spacecraft coordinate system are colinear with the principal axes of the vehicle. In that case, the moment-of-inertia tensor has only three components. The nondiagonal elements of the moment-of-inertia tensor vanish in this case so that equation (9.24) becomes

$$\begin{pmatrix} L_1\hat{s}_1 \\ L_2\hat{s}_2 \\ L_3\hat{s}_3 \end{pmatrix} = \begin{pmatrix} I_{11} & 0 & 0 \\ 0 & I_{22} & 0 \\ 0 & 0 & I_{33} \end{pmatrix} \begin{pmatrix} \omega_1(s,\,i)\hat{s}_1 \\ \omega_2(s,\,i)\hat{s}_2 \\ \omega_3(s,\,i)\hat{s}_3 \end{pmatrix} \tag{9.25}$$

which can be rewritten as an ordinary vector equation in the spacecraft centered s coordinate system:

$$\mathbf{L} = L_1\hat{s}_1 + L_2\hat{s}_2 + L_3\hat{s}_3 = I_{11}\omega_1(s,\,i)\hat{s}_1 + I_{22}\omega_2(s,\,i)\hat{s}_2 + I_{33}\omega_3(s,\,i)\hat{s}_3 \tag{9.26}$$

We are now ready to derive the Euler equations, which are the equations of motion of a rigid body. They are, of course, based on Newton's second law of motion relating the applied torque to the time rate of change of the angular momentum:

$$\mathbf{T} = \frac{d\mathbf{L}}{dt}(i) = \frac{d\mathbf{L}}{dt}(s) + \boldsymbol{\omega}(s,\,i) \times \mathbf{L} \tag{9.27}$$

In order to obtain equation (9.27), we used the general relationship between the time derivative of a vector in the inertial system i and in the spacecraft system s defined in equation (9.12). Remembering the relationship between the angular momentum and the angular velocity (9.21), we can write

$$\mathbf{T} = \frac{dI\,\boldsymbol{\omega}}{dt}(i) = \frac{dI\,\boldsymbol{\omega}}{dt}(s) + \boldsymbol{\omega}(s,\,i) \times I\,\boldsymbol{\omega} \tag{9.28}$$

Since the s system is centered on the spacecraft, the angular velocity $\boldsymbol{\omega}$ of the spacecraft is equal to the angular velocity of the s coordinate system with respect to the i system, $\boldsymbol{\omega}(s,\,i)$. We can therefore write

$$\boldsymbol{\omega}(s,\,i) = \boldsymbol{\omega} = \omega_1\hat{s}_1 + \omega_2\hat{s}_2 + \omega_3\hat{s}_3 \tag{9.29}$$

and therefore the torque is

$$\mathbf{T} = \frac{dI\,\boldsymbol{\omega}}{dt}(i) = \frac{dI\,\boldsymbol{\omega}}{dt}(s) + \boldsymbol{\omega} + I\,\boldsymbol{\omega} \tag{9.30}$$

The first term on the right side of equation (9.30) can be written as follows:

$$\frac{dI\,\boldsymbol{\omega}}{dt}(s) = \frac{d}{dt}\begin{bmatrix} I_{11} & 0 & 0 \\ 0 & I_{22} & 0 \\ 0 & 0 & I_{33} \end{bmatrix}\begin{bmatrix} \omega_1\hat{s}_1 \\ \omega_2\hat{s}_2 \\ \omega_3\hat{s}_3 \end{bmatrix} \tag{9.31}$$

$$= I_{11}\frac{d\omega_1}{dt}\hat{s}_1 + I_{22}\frac{d\omega_2}{dt}\hat{s}_2 + I_{33}\frac{d\omega_3}{dt}\hat{s}_3$$

since the components of the moment of inertia are constant in the s system.

The second term is a vector product of $\boldsymbol{\omega}$ and $I\boldsymbol{\omega}$, which can be evaluated as follows:

$$\begin{bmatrix} \omega_1\hat{s}_1 \\ \omega_2\hat{s}_2 \\ \omega_3\hat{s}_3 \end{bmatrix} \times \begin{bmatrix} I_{11} & 0 & 0 \\ 0 & I_{22} & 0 \\ 0 & 0 & I_{33} \end{bmatrix}\begin{bmatrix} \omega_1\hat{s}_1 \\ \omega_2\hat{s}_2 \\ \omega_3\hat{s}_3 \end{bmatrix}$$

$$= \begin{bmatrix} \omega_1\hat{s}_1 \\ \omega_2\hat{s}_2 \\ \omega_3\hat{s}_3 \end{bmatrix} \times \begin{bmatrix} I_{11}\omega_1\hat{s}_1 \\ I_{22}\omega\hat{s}_2 \\ I_{33}\omega_3\hat{s}_3 \end{bmatrix}$$

$$= \omega_1\hat{s}_1 \times I_{11}\omega_1\hat{s}_1 + \omega_1\hat{s}_1 \times I_{22}\omega_2\hat{s}_2 + \omega_1\hat{s}_1 \times I_{33}\omega_3\hat{s}_3$$

$$+ \omega_2\hat{s}_2 \times I_{11}\omega_1\hat{s}_1 + \omega_2\hat{s}_2 \times I_{22}\omega_2\hat{s}_2 + \omega_2\hat{s}_2 \times I_{33}\omega_3\hat{s}_3$$

$$+ \omega_3\hat{s}_3 \times I_{11}\omega_1\hat{s}_1 + \omega_3\hat{s}_3 \times I_{22}\omega_2\hat{s}_2 + \omega_3\hat{s}_3 \times I_{33}\omega_3\hat{s}_3$$

$$= \hat{s}_1(I_{33} - I_{22})\,\omega_2\omega_3 + \hat{s}_2(I_{11} - I_{33})\,\omega_1\omega_3$$

$$+ \hat{s}_3(I_{22} - I_{11})\omega_1\omega_2$$

where the relationships (9.9) and (9.10) have been applied. We now have the following expressions for the components of the torque vector:

$$T_1 = I_{11}\frac{d\omega_1}{dt} + (I_{33} - I_{22})\omega_2\omega_3 \qquad (9.32)$$

$$T_2 = I_{22}\frac{d\omega_2}{dt} + (I_{11} - I_{33})\omega_1\omega_3 \qquad (9.33)$$

$$T_3 = I_{33}\frac{d\omega_3}{dt} + (I_{22} - I_{11})\omega_1\omega_2 \qquad (9.34)$$

The relationships (9.32), (9.33), and (9.34) are called the Euler equations, and these relate the components of the torque vector in the i system to the components of the moment of inertia in the s system and the angular velocities. Thus, the Euler equations can be used to calculate the motion of the spacecraft when the torques are known.

The conventional parameters that relate the inertial coordinate system i to the spacecraft system s are called the Euler angles. They are illustrated in Figure 9.2. The best way to see how the Euler angles can be constructed is to start with Figure 9.1 and to use the following procedures:

1. Reorient the two-dimensional coordinate system of Figure 9.1 in such a way that i_3 axis is vertical, as shown in Figure 9.2. This is achieved by first rotating the system clockwise around the i_1 axis to 90° and, once the system is in that orientation, rotating it clockwise through 90° around the i_3 axis. Once this is accomplished, the inertial i system is oriented as shown in Figure 9.2.
2. Rotate the spacecraft s system around the i_3 axis through the angle φ.
3. Rotate the spacecraft s system around the axis $\hat{n}$ by the angle θ. The unit vector $\hat{n}$ is perpendicular to the plane defined by r and $\hat{s}_3$.
4. Rotate the spacecraft s system around the $\hat{s}_3$ axis through the angle ψ.

These angles conventionally define the orientation of a rigid body in space. If we consider a spinning top, then rotation around $\hat{s}_3$ determines the spin of the top with an angular velocity of $d\psi/dt$, rotation around the i_3 axis determines the precession of the top with an angular velocity $d\varphi/dt$, and rotation around the $\hat{n}$ axis, which changes the angle θ, is $d\theta/dt$ and is called nutation.

The result of these operations yields the picture shown in Figure 9.2. What we want to do now is to solve the Euler equations for a spacecraft

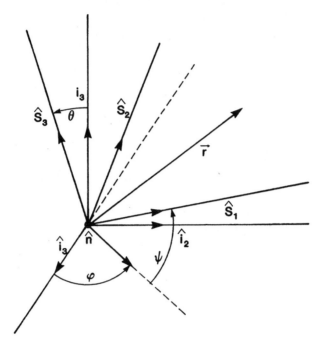

FIGURE 9.2

that has no external forces acting upon it. We will then show how the spacecraft responds to various external torques.

In order to do this, we first have to rewrite the Euler equations in terms of the Euler angles defined in Figure 9.2. The angular velocity of the spacecraft is defined by the angular velocity of the s system with respect to the inertial i system. These are defined in equation (9.29). We can also write the angular velocity of the spacecraft in terms of the rates of change of the Euler angles:

$$\boldsymbol{\omega} = \frac{d\varphi}{dt}\,\hat{i}_3 + \frac{d\theta}{dt}\,\hat{n} + \frac{d\psi}{dt}\,\hat{s}_3 \tag{9.35}$$

In order to solve the Euler equations in terms of the Euler angles, we need to make a transformation of the unit vector basis set of equation (9.35) to the basis set $\hat{s}_1$, $\hat{s}_2$, and $\hat{s}_3$. The unit vector $\hat{n}$ has components along $\hat{s}_1$ and $\hat{s}_2$ as follows:

$$\hat{n} = \hat{s}_1\cos\psi + \hat{s}_2\sin\psi \tag{9.36}$$

In the case of the angular velocity component around the $\hat{i}_3$ axis, the precession, we can write in terms of components along the $\hat{s}_3$ axis and the $\mathbf{r}$ vector as follows:

$$\hat{i}_3 \frac{d\varphi}{dt} = \hat{s}_3 \frac{d\varphi}{dt} \cos\theta + \mathbf{r} \frac{d\varphi}{dt} \sin\theta \qquad (9.37)$$

The vector $\mathbf{r}$ has the following components along $\hat{s}_1$ and $\hat{s}_2$:

$$\mathbf{r} = \hat{s}_1 \sin\psi + \hat{s}_2 \cos\psi \qquad (9.38)$$

We can now write down the components of the vector $\boldsymbol{\omega}$ along $\hat{s}_1$, $\hat{s}_2$, and $\hat{s}_3$ by comparing equations (9.38), (9.37), (9.36), and (9.35) with equation (9.29). The result is

$$\omega_1 = \frac{d\varphi}{dt} \sin\theta \sin\psi + \frac{d\theta}{dt} \cos\psi$$

$$\omega_2 = \frac{d\varphi}{dt} \sin\theta \cos\psi - \frac{d\theta}{dt} \sin\psi \qquad (9.39)$$

$$\omega_3 = \frac{d\psi}{dt} + \frac{d\varphi}{dt} \cos\theta$$

These equations are called the Euler rate equations, and we can use them to get solutions of the Euler equations for cases for practical interest.

The first case we shall consider is that of the spin-stabilized spacecraft. Spin stabilization is most commonly used for communications satellites placed in geosynchronous orbit. Spin stabilization is desirable in this case because the satellites are required to have long lifetimes. This means that the use of small jets or rockets to control the spacecraft needs to be minimized to save fuel or working fluids. Furthermore, communications satellites do not generally require large orientation changes so that the methods for changing spin orientation that will be developed are adequate. Communications satellites are cylindrical in shape and have their spin axis perpendicular to Earth's equatorial plane. Thus, it is advantageous to place the solar cells that provide electrical power on the cylindrical surface of the spacecraft. This arrangement has the advantage that is provides a constant power level for the operation of the spacecraft. The antenna on the spacecraft must always be pointed toward the same spot on the surface of Earth. Therefore, the antenna cannot rotate and must be mounted on what is called a "despun" platform attached to a bearing with a shaft that remains stationary while the spacecraft rotates. A picture of a typical communications satellite that operates on these principles, the *IN-TELSAT IV-A,* is shown in Figure 9.3.

Another important class of spin-stabilized spacecraft is comprised of those designed to fly to the outer reaches of the solar system. An example of this is the *Pioneer 10* spacecraft shown in Figure 10.13. *Pioneer 10* achieved a number of "firsts," including the flyby of Jupiter. In 1990, *Pioneer 10* passed beyond the orbit of Pluto and therefore became the first man-made object to leave the solar system. The design of the *Pioneer 10*

FIGURE 9.3 This is a picture of *INTELSAT IV-A,* a spin-stabilized geosynchronous communications satellite. The body of the satellite spins and the antenna is mounted on a "despun" platform so that it always points at the same spot on the surface of the Earth.

is dominated by the high-gain dish antenna necessary for communication over long distances. Spacecraft designed to reach the outer planets cannot use solar power because the Sun is too weak. *Pioneer 10* was powered by two radioisotope thermal power supplies mounted on booms, as shown in Figure 10.13. A third boom carried a magnetometer to balance the two power supplies when the spacecraft rotates around its principal axis. This axis runs through the center of the disk antenna and is perpendicular to the plane of the dish. In operation, the spin axis of *Pioneer 10* lies in the plane of the ecliptic. The spinning spacecraft is then maneuvered in such a way that the principal axis, and hence the antenna, always points toward the Earth.

Having described the properties of spin-stabilized spacecraft, we are now ready to develop the mathematics that governs their control. All spin-stabilized spacecraft are designed to be axially symmetric. This means that they have one prinicpal axis around which they rotate and two other axes that have equal moments of inertia. Thus, if the prinicipal axis is $\hat{s}_3$, with a principal moment of inertia, I_{33}, we can write:

$$I_{11} = I_{22} \tag{9.40}$$

We can now write the Euler equations for the axi symmetric spacecraft with no external forces acting on it:

$$T_1 = 0 = I_{11}\frac{d\omega_1}{dt} + (I_{33} - I_{11})\omega_2\omega_3 \tag{9.41}$$

$$T_2 = 0 = I_{22}\frac{d\omega_2}{dt} + (I_{11} - I_{33})\omega_1\omega_3 \tag{9.42}$$

$$T_3 = 0 = I_{33}\frac{d\omega_3}{dt} + (I_{11} - I_{11})\omega_2\omega_3 \tag{9.43}$$

Equation (9.43) can be easily solved because:

$$I_3\frac{d\omega_3}{dt} = 0 \tag{9.44}$$

therefore

$$\omega_3 = \Omega = \text{constant} \tag{9.45}$$

which says that the spin rate of the satellite around the principal axis $\hat{s}_3$ is equal to the constant Ω. This result is to be expected because there are no external torques. The other two Euler equations are

$$\frac{d\omega_1}{dt} = -\frac{I_{33} - I_{11}}{I_{11}} \omega_2 \Omega \tag{9.46}$$

and

$$\frac{d\omega_2}{dt} = \frac{I_{33} - I_{11}}{I_{11}} \omega_1 \Omega \tag{9.47}$$

these are two coupled first-order differential equations that can be solved by reformulating them into a single second-order equation by differentiating equation (9.46) with respect to time and then using equation (9.47):

$$\frac{d^2\omega_1}{dt^2} = -\alpha^2 \omega_1 \tag{9.48}$$

where

$$\alpha = \frac{I_{33} - I_{11}}{I_{11}} \Omega \tag{9.49}$$

The solution of equation (9.48) is the same as that of a simple harmonic oscillator:

$$\omega_1(t) = \omega_0 \sin \alpha(t - t_0) \tag{9.50}$$

$$\omega_2(t) = -\omega_0 \cos \alpha(t - t_0) \tag{9.51}$$

Now, we will use the Euler rate equations to find the expressions for the spin rate $d\Psi/dt$ and the precession rate $d\varphi/dt$. We can arbitrarily assume that the angle θ is constant, and we will shortly understand the role that θ plays in the motion of the spacecraft. Using the Euler rate equations and the solutions (9.50) and (9.51), we have

$$\omega_1(t) = \omega_0 \sin \alpha t = \frac{d\varphi}{dt} \sin \theta \sin \psi \tag{9.52}$$

$$\omega_2(t) = -\omega_0 \cos \alpha t = \frac{d\varphi}{dt} \sin \theta \cos \psi \qquad (9.53)$$

$$\omega_3(t) = \Omega = d\psi \backslash dt + \frac{d\varphi}{dt} \cos \theta \qquad (9.54)$$

Squaring and adding equations (9.52) and (9.53) yield

$$\omega_0^2 = \left(\frac{d\varphi}{dt} \right)^2 \sin^2 \theta \qquad (9.55)$$

and taking the square root gives

$$\omega_0 = \frac{d\varphi}{dt} \sin \theta \qquad (9.56)$$

Comparing equation (9.53) with equation (9.56) yields

$$-\cos \alpha t = \cos \psi \qquad (9.57)$$

or

$$-\alpha t = \psi \qquad (9.58)$$

Solving equation (9.54) for $d\varphi/dt$ yields

$$\frac{d\varphi}{dt} = \frac{\Omega - d\psi/dt}{\cos \theta} \qquad (9.59)$$

solving the expression in equation (9.49) for Ω gives

$$\Omega = \frac{I_{11}}{I_{33} - I_{11}} \alpha \qquad (9.60)$$

and from equation (9.58) we have

$$\frac{d\psi}{dt} = -\alpha \qquad (9.61)$$

so that we can write

$$\Omega = -\frac{I_{11}}{I_{33} - I_{11}} \frac{d\psi}{dt} \tag{9.62}$$

which, when substituted into equation (9.59) and performing some algebraic manipulation, yields the following result:

$$\frac{d\varphi}{dt} = \frac{I_{33}}{(I_{11} - I_{33})\cos\theta} \frac{d\psi}{dt} \tag{9.63}$$

Equation (9.63) relates the spin rate of the satellite $d\psi/dt$ to the precession rate $d\varphi/dt$ as a function of the moments of inertia and the angle θ. It is this formula that permits us, under certain restricted conditions, to calculate quantitatively how the spin-stabilized spacecraft behaves when outside torques are applied.

We are now ready to solve the problem of changing the orientation of a spin-stabilized spacecraft. To do this, refer to Figure 9.4, which illustrates the precession of a disk-shaped spacecraft with a radius R. The thrusters are located on the rim of the disk. They each supply a thrust **F** as shown. In order to move the angular momentum vector through the angle θ, a torque must be applied so that an angular momentum change Δ**L** is added to the initial angular momentum **L**$_1$. In Figure 9.4, the disk represents the spacecraft, and there are two small jets mounted on the rim of the disk that supply the torque. If the radius of the spacecraft is R and the thrust of the jet is F, then the torque vector is

$$\mathbf{T} = 2(\mathbf{R} \times \mathbf{F}) \tag{9.64}$$

The vector product defined in equation (9.64) provides a change in the initial angular momentum vector **L**$_1$ by Δ**L**, which is a vector perpendicular to **L**$_1$, as shown in Figure 9.4. The angular momentum change is

$$\Delta\mathbf{L} = \mathbf{T}\,\Delta\tau \tag{9.65}$$

where $\Delta\tau$ is the time interval over which the torque is applied. This equation is derived from equation (9.15). The new angular momentum vector of the spacecraft points in the direction indicated that makes an angle $\theta/2$ defined by the relationship

$$\tan\frac{\theta}{2} = \frac{|\Delta\mathbf{L}|}{|\mathbf{L}_1|} \tag{9.66}$$

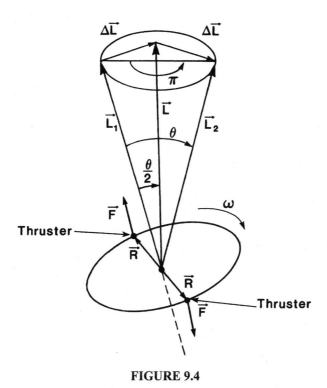

FIGURE 9.4

The vector $\mathbf{L}_1$, in addition, precesses around the direction of the new angular momentum vector $\mathbf{L}$. It is very important to recognize that the state of affairs described here only works if the angular momentum change vector $\Delta\mathbf{L}$ is small compared to the vector $\mathbf{L}_1$. If this condition is not fulfilled, then motion around the vector $\hat{n}$, that is, the angular velocity component $d\theta/dt$, is no longer zero and the whole analysis presented here breaks down. From Figure 9.4, it can be seen that if the spacecraft is permitted to precess through an angle of 180° (π), and if the same torque that initiated the precession is now applied in the indicated direction, the precession will stop and the spacecraft will then spin around an axis that makes an angle θ with the original spin axis. The new angular momentum $\mathbf{L}_2$ will have the same magnitude as the original angular momentum $\mathbf{L}_1$ subject to the condition on $|\Delta\mathbf{L}|$ with respect to $|\mathbf{L}_1|$ that has already been mentioned.

A relationship can now be developed that will yield the thrust of the steering jet, F, necessary to move the spin vector of the spacecraft through an angle θ in terms of the moments of inertia and the spin rate of the spacecraft. Using equation (9.63), we first require that the angle $d\varphi$ be

$$d\varphi = \pi \qquad (9.67)$$

The time interval over which the force must be applied is

$$dt = \Delta\tau = \frac{\pi(I_{11} - I_{33})}{I_3\Omega} \cos\frac{\theta}{2} \qquad (9.68)$$

The magnitude of the thrust F that must be supplied by the jet is then given by the relationship that can be recognized by looking at Figure 9.4:

$$2R|\mathbf{F}|\,\Delta\tau = |\Delta\mathbf{L}| = |\mathbf{L}_1|\tan\frac{\theta}{2} \qquad (9.69)$$

so that

$$|\mathbf{F}| = \frac{|\mathbf{L}_1|}{2R}\frac{I_3\Omega}{\pi(I_{11} - I_{33})\cos(\theta/2)}\tan\frac{\theta}{2} \qquad (9.70)$$

We have already mentioned the fact that $\Delta\mathbf{L}$ must be small compared to $\mathbf{L}_1$ in order for equation (9.70) to work. This means, of course, that only small changes in θ can be accurately represented by equation (9.70). In fact, if we tried to apply this formula to making a change in orientation of the spacecraft by 180°, a complete reversal of the direction of the angular momentum vector, an infinite thrust would be required. This is obviously not reasonable and reflects the nature of the approximation that we have discussed. In practice, large changes in the orientation of spin-stabilized spacecraft are executed by making a successive number of small changes, as illustrated in Figure 9.5. In the figure, the orientation of the spacecraft spin vector is changed by an angle of 90° by making five successive changes of 18°, each of which can be achieved by a relatively small application of thrust.

The second method of stabilizing spacecraft that is in common use is called *three-axis stabilization*. Spin-stabilized spacecraft rely on the *gyroscopic effect,* which means that the angular momentum vector of a rotating body maintains its direction in space unless some outside torque is applied. If that angular momentum has a large enough magnitude, then it is also hard to change the angular momentum, and hence the spacecraft is

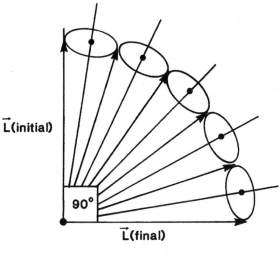

$\vec{L}$(initial)

90°

$\vec{L}$(final)

FIGURE 9.5

"stabilized." Three axis stabilization depends on equipping a spacecraft with the means for orienting the spacecraft accurately with respect to a fixed external axis that is usually determined by some kind of a star sensor. This means of stabilization does not depend on inertial effects but rather on an active control system that maintains the direction in space accurately by a feedback control system locked to the axis defined by the star sensor. Three-axis stabilization is essential if accuracy of pointing is required. For example, the Hubble Space Telescope has a very sophisticated three-axis pointing system because accuracy is most important for astronomy. The two Voyager spacecraft that conducted the historic missions to all of the outer planets except Pluto had a three-axis stabilization system that permitted the Voyager spacecraft to produce spectacular high-resolution pictures of Jupiter and its satellites, Saturn, Titan, and the rings, and the Uranus and Neptune systems. *Landsat* and many of the classified Earth observation systems have three-axis stabilization in order to obtain high-resolution pictures of Earth's surface. Finally, the effort to put landers on the planet Mars during the Viking program also required three-axis stabilization.

The simplest way to achieve three-axis stabilization is shown in Figure 9.6. A highly redundant set of thruster pairs mounted on each edge, as shown on the notional satellite in Figure 9.6, supply the force couples, or torques, necessary to rotate the spacecraft. This is a very simple system,

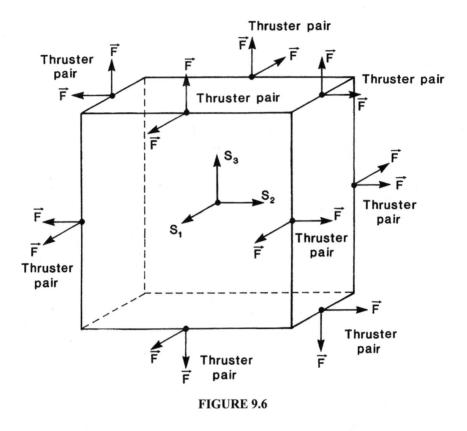

FIGURE 9.6

and it is easy to see that any orientation can be achieved by the appropriate application of torques. The forces required to move the spacecraft are relatively small even for a large spacecraft. In many instances, small compressed gas jets rather than rockets can be used. The best example of three-axis stabilization using this method is the space shuttle. The attitude of the space shuttle when it is in Earth orbit is controlled by several sets of small rockets using hypergolic fuels (hydrazine and nitrous oxide in this case). These thrusters are part of what is called the Reaction Control System (RCS). There is one forward module with 14 primary thrusters each with 870 lb thrust and two vernier thrusters of 25 lb each. There are two aft RCS pods, each containing 12 primary thrusters (870 lb) and two vernier thrusters (25 lb). All of these rockets can be turned on and off at will. It is remarkable that these relatively low thrust devices can accurately control the attitude of a vehicle that weighs 200,000 lb (90,910 kgm).

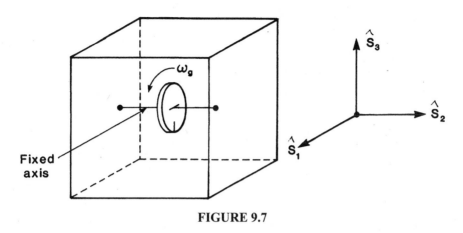

FIGURE 9.7

The principal drawback of using rockets or jets to achieve three-axis stabilization is that they require fuel or compressed gas to operate. Therefore, the system has a limited lifetime because it will not work when the fuel of the rocket or the working fluid of the jet is exhausted.

An alternate way of changing the orientation of a spacecraft is to use a momentum wheel or a gyroscope to rotate the vehicle and to reorient it. The principal advantage of momentum wheels or gyroscopes is that they work as long as electric power is available to keep the wheels turning. A simple example of a momentum wheel is shown in Figure 9.7. The axis of the momentum wheel is fixed in the spacecraft and it is spun by a small electric motor that is powered by electricity obtained from solar panels mounted on the spacecraft. In the case shown, the momentum wheel is mounted parallel to the $\hat{s}_2$ axis. A simple maneuver is to rotate the spacecraft. The total angular momentum is always zero at the beginning of the procedure, and the angular momentum of the wheel is $\mathbf{L}_g$. When the gyroscope is started, there is a transfer of angular momentum to the spacecraft given by the following relationships:

$$\mathbf{L}_T = \mathbf{L}_g + \mathbf{L}_s = 0 \tag{9.71}$$

and for each of the angular momenta we have

$$\mathbf{L}_g = I_g \boldsymbol{\omega}_g \tag{9.72}$$

and

$$\mathbf{L}_s = I_s \boldsymbol{\omega}_s \tag{9.73}$$

Since all of the vectors in the above equations are assumed to be colinear, we can use their scalar values:

$$I_g(\omega_g - \omega_s) - I_s \omega_s = 0 \tag{9.74}$$

so that the final angular velocity of the spacecraft is

$$\omega_s = \frac{I_g \omega_g}{I_g + I_s} \tag{9.75}$$

where ω_s is the angular velocity of the spacecraft after the gyroscope is spun up to an angular velocity of ω_g and I_g and I_s are the respective moments of inertia of the spacecraft and the momentum wheel.

If "momentum wheels" are mounted on the spacecraft along the three principal axes of the vehicle, then they can be used to produce rotations of the spacecraft around the other two axes of the spacecraft by a similar method. Thus, the spacecraft can be moved to point along an arbitrary axis by techniques similar to those described for the spin-stabilized spacecraft. A difficulty with the use of momentum wheels to point spacecraft is that the moment of inertia of the spacecraft is usually much larger than that of the momentum wheel ($I_s \gg I_g$). Thus, successive pointing maneuvers may sometimes require that the momentum wheels be spun to very high angular velocities. Eventually, this requires that the spacecraft at certain times be stabilized with jets or rockets and that the momentum wheels be stopped. This maneuver is called a "momentum dump."

Another way of controlling the orientation of a spacecraft is to use momentum wheels but mounted on gimbals rather than on a fixed axis in the spacecraft. In this configuration, the system is called a *control moment gyroscope*. Control moment gyros are used to point the Hubble Space Telescope, Skylab, and certain Earth observation satellites. All of these spacecraft have to be rapidly and accurately pointed on a demanding schedule. Figure 9.8 illustrates the operation of a control moment gyroscope. An angular momentum can be imposed on the spacecraft by applying a torque along the gimbal axis of other gyroscope, changing its angular momentum by $\Delta \mathbf{L}_g$. Since no external torque is applied, there will be an equal and opposite reaction of the spacecraft. Therefore

$$\Delta \mathbf{L}_g = -\Delta \mathbf{L}_s \tag{9.76}$$

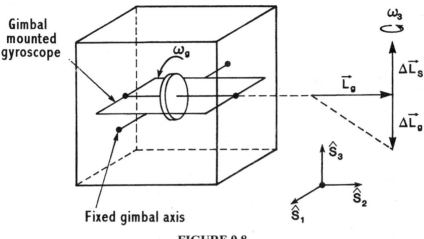

FIGURE 9.8

The spacecraft will not acquire an angular velocity of ω_s around the $\hat{s}_3$ axis:

$$\omega_3 = \frac{I_g \omega_g \sin \theta}{I_s} \qquad (9.77)$$

The spacecraft rotation can be stopped by the application of an equal but opposite torque. This illustrates how a gyroscope can be used to change the orientation of the spacecraft. The principle described can be expanded to provide any arbitrary orientation of the spacecraft by mounting three gyroscopes on the spacecraft pointing along the three mutually perpendicular axes $\hat{s}_1$, $\hat{s}_2$, and $\hat{s}_3$ of the spacecraft-based coordinate system.

A number of good books are available that provide further reading on the general subject of the mechanics of solid bodies and spacecraft control systems. An excellent general treatment of solid-body dynamics at the undergraduate level is *Mechanics* by Keith R. Symon. There are several excellent books on spacecraft dynamics and maneuvering systems. Among these are *Spaceflight Dynamics* by William E. Wiesel, *Modern Spacecraft Dynamics and Control* by Marshall H. Kaplan and *Spacecraft Attitude Determination and Control,* edited by James R. Wertz. All of these works are recommended to the reader for further study.

PROBLEMS

9.1. A communications satellite is designed as a right circular cylinder with moments of inertia $I_{11} = I_{22}$ and I_{33}, the last mentioned being the moment of inertia around the axis of symmetry. Also, $I_{11} = I_{22} = 10\,I_{33}$. The satellite is equipped with thrusters designed to supply an angular momentum vector ΔL perpendicular to the angular momentum L_3 of the satellite as it rotates around the axis of symmetry where $|L_3| = 100|\Delta L|$. The satellite is spin stabilized by rotation around the axis of symmetry at the rate of 6 revolutions per minute. A maneuver that rotates the satellite's spin axis (the axis of symmetry) through 90° is initiated using the thrusters in a series of steps that requires precessing the axis of the satellite through a series of small angles defined by equation (9.66) in the text.

(a) How many steps are necessary to execute the maneuver?
(b) How long does the maneuver take to execute?

9.2. The *Pioneer Venus* orbiter spacecraft was a right circular solid cylinder of mass M and radius R. The orbiter was designed to be spin stabilized by rotation around the axis of symmetry. The orbiter's rotation around this axis is initiated by a momentum wheel mounted on the axis of symmetry. The momentum wheel has a mass, m, and a radius, r, and it is designed so that its mass is concentrated on the rim with spokes of negligible weight compared to the rim. The following were the relationships between the masses and the radii:

$$M = 100\, m$$

$$R = 10\, r$$

The spin rate required to stabilize the *Pioneer Venus* orbiter was 10 revolutions per minute. What spin rate was necessary for the momentum wheel to achieve the spin rate required for the orbiter?

9.3. The attitude of the Hubble Space Telescope is determined by a control moment gyroscope. (This is essentially a momentum wheel mounted on a gimbal.) The gyroscope has a mass, m, a radius, r, and an angular velocity ω_g. It is designed in the same way as the momentum wheel in the previous problem. The gyroscope is mounted on the principal axis of the telescope. The original angular momentum of the system is equal to $mr^2\omega_g$. The angular momentum of the gyroscope pointing along the principal axis of the telescope. It is now desired to reorient the telescope using the control moment gyroscope. To do this, a torque is applied to the gimbal of the gyroscope around an axis perpendicular to the gyroscope spin axis. The added angular momentum rotates the angular momentum vector of the gyroscope through an angle θ which is compensated by a rotation of the whole spacecraft along an axis perpendicular both to the principal axis of the telescope and to the gimbal axis of the gyroscope. This happens because the angular momentum of the entire spacecraft cannot change.

(a) What is the angular velocity of the space telescope if the angle θ is 5.7% (1/10 of a radian), the ratio of the moment of inertia of the gyroscope to that of the space telescope around the rotation axis is 10^{-5} and the angular velocity of the gyroscope is 1000 revolutions per minute?

(b) At what time after the initial torque is applied to the gyroscope gimbal must an equal and opposite torque be applied so that the principal axis of the telescope is rotated through 90°?

CHAPTER 10

PLANETARY EXPLORATION

The solar system consists of the Sun, the planets and their satellite or ring systems, the asteroids, and the comets. The Sun is at the center of the solar system and the nine planets, in the order Mercury, Venus, Earth, Mars, Jupiter, Saturn, Uranus, Neptune, and Pluto, move in orbits around the Sun. The Sun contains about 99.86% of the mass of the solar system, which permits treatment of the motion of each planet in the solar system as a two-body problem to a first approximation. The orbits of the planets (with the exception of Pluto) lie very nearly in the same plane, which is called the ecliptic plane.

The planets can be divided into two distinct classes, the inner (or "terrestrial") planets, that is, Mercury, Venus, Earth, and Mars, and the outer (or "giant") planets, Jupiter, Saturn, Uranus, and Neptune. The terrestrial planets are small, a few thousand kilometers in diameter, and have a mean density of about 5 g/cm^3, which is the same as silicate rock. Earth is the largest of the terrestrial planets. The outer planets are large, having masses between 15 and 300 times that of Earth and diameters between 4 and 11 times that of Earth. The mean density of the outer planets is about 1.4 g/cm^3, which is about 40% larger than that of water. The ninth planet, Pluto, is a special case, since it does not belong to either class. Pluto is smaller than Mercury and probably has roughly the same density as the terrestrial planets. Pluto's orbit is both the most eccentric of all planetary orbits, and it also lies in a plane which makes the largest angle with the ecliptic plane. The properties of the solar system are shown in Table 10.1.

188

TABLE 10.1. The Planets

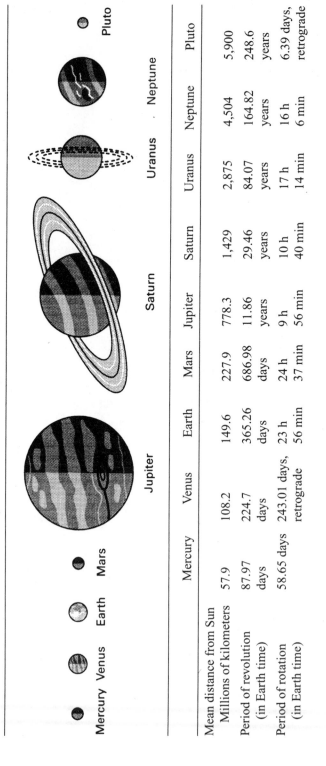

	Mercury	Venus	Earth	Mars	Jupiter	Saturn	Uranus	Neptune	Pluto
Mean distance from Sun Millions of kilometers	57.9	108.2	149.6	227.9	778.3	1,429	2,875	4,504	5,900
Period of revolution (in Earth time)	87.97 days	224.7 days	365.26 days	686.98 days	11.86 years	29.46 years	84.07 years	164.82 years	248.6 years
Period of rotation (in Earth time)	58.65 days	243.01 days, retrograde	23 h 56 min	24 h 37 min	9 h 56 min	10 h 40 min	17 h 14 min	16 h 6 min	6.39 days, retrograde

Inclination of axis (degrees)	0.0	177.3	23.5	25.2	3.08	26.7	97.9	29.6	122
Inclination of orbit to ecliptic (degrees)	7.0	3.39	0.0	1.85	1.31	2.49	0.77	1.77	17.15
Eccentricity (degrees)	0.206	0.007	0.017	0.093	0.048	0.056	0.046	0.010	0.248
Equatorial diameter Kilometers	4,878	12,104	12,756	6,784	142,796	120,000	50,800	48,600	~2,500
Atmosphere	Essentially none	Carbon dioxide	Nitrogen, oxygen	Carbon dioxide	Hydrogen, helium	Hydrogen, helium	Hydrogen, helium	Hydrogen, helium	Methane
Satellites	None	None	1	2	16	18	15	8	1
Rings	None	None	None	None	1	Thousands	11	5	Probably none
Mean density g/cm³	5.0	4.9	5.52	4.2	1.33	0.71	1.26	1.61	—

Source: NASA Pocket Statistics, 1996.

189

The asteroids are small bodies, usually with diameters less than 100 kilometers, that lie essentially between the orbits of Mars and Jupiter. Their densities tend to be in the range of the terrestrial planets. Most asteroids move in relatively circular orbits, but some are in highly eccentric orbits that actually cross the orbit of Earth. We will return to the problem of understanding why such "Earth-crossing" asteroids exist in Chapter 13. Asteroids are thought to be the remnants of a planet that once existed between the orbits of Mars and Jupiter. This planet apparently disintegrated for some unknown reason, giving rise to the asteroids. A typical large asteroid, Ceres, has a diameter of about 770 km. It moves in a roughly circular orbit with a diameter of about 2.7 times that of Earth, and the orbital plane has a relatively high inclination (10.4°) to the ecliptic plane. An Earth-crossing asteroid might be typified by Eros, which has a diameter of about 20 km, and its orbit has a large eccentricity (0.22) and also an angle of inclination to the ecliptic of about 10.4°.

The material from which comets originated is located in the outer reaches of the solar system, called the *Oort cloud,* after the Dutch astronomer J. H. Oort, who first identified it. Comets are formed by the accretion of ice particles in a region beyond the orbit of Pluto. As is the case with asteroids, most of these bodies move in roughly circular orbits. However, sometimes these ice bodies can be deflected into orbits that are highly eccentric by the same mechanisms that operate on the asteroids. The "ice bodies" that assume such orbits become "comets" as we observe them periodically. Some comets that have been observed are actually in orbits that have positive energies and will therefore never return.

Most of the planets in the solar system have satellites. By far the largest satellite in the solar system *relative to its planet* is the Moon, which has about 1.2% of the mass of Earth. Some have even called the Earth–Moon system a "double planet." Jupiter and Saturn both have satellites that are larger than Earth's Moon but are very much smaller than Jupiter and Saturn. Jupiter has four large satellites, Io, Europa, Ganymede, and Callisto, which were discovered in 1610 by Galileo, when he first looked at Jupiter through his telescope. Hence, these are called the "Galilean" satellites. Jupiter also has a large number of smaller satellites. The largest satellite in the solar system is Titan, which revolves around Saturn. Titan is about the same size as the planet Mercury, or about 5% of the mass of Earth. Saturn and all of the outer planets except Pluto have ring systems, which we shall discuss later in this chapter.

The solar system probably originated when a rotating gas cloud collapsed because of the mutual gravitational attraction of the molecules in the cloud. As the rotating cloud collapses, the temperature of the gas, which is mostly hydrogen, rises. Eventually, the gas at the center gets hot enough so that nuclear reactions can occur, and the center becomes a star, or a "sun" in the case of the solar system. The rotating motion of the gas cloud while this process occurs is turbulent because as the cloud collapses and the angular velocity increases, the average Reynolds number becomes larger and eventually exceeds the critical value at which laminar flow is possible. Thus, we can imagine a situation, such as the one shown in Figure 10.1, in which a large number of turbulent eddies have developed in the gas cloud. It is thought that these eddies are the "protoplanets" from which the planets of the solar system have evolved. Such gas clouds in which protoplanets develop have now been observed by the Hubble Space Telescope in the great nebula that is located in the constellation Orion.

In addition to the formation of turbulent eddies, the rotating gas cloud also ultimately collapses into a rotating disk, as shown in Figure 10.2. This collapse into a disk occurs because the motion of the cloud is dissipative so that energy is lost. However, no torques from the outside of the system exist so that the cloud tries to maximize the angular momentum with respect to the energy contained in the rotation. Since both the mass of the cloud and the angular momentum stay roughly constant during the

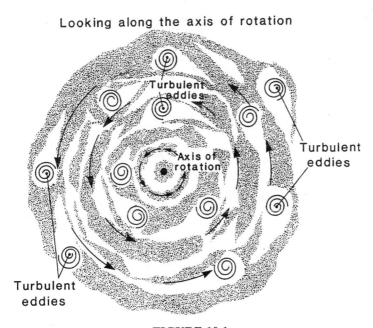

FIGURE 10.1

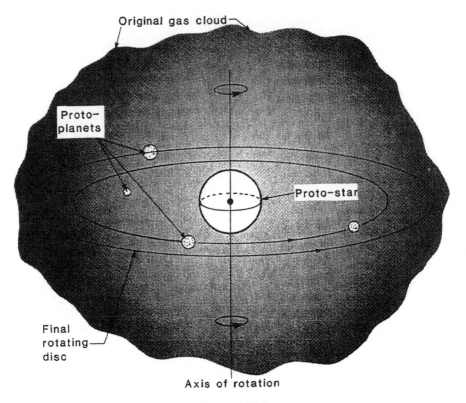

Original gas cloud

Proto-planets

Proto-star

Final rotating disc

Axis of rotation

FIGURE 10.2

collapse, the only way to accomplish this is to maximize the moment of inertia of the system. For a given mass, the moment of inertia of a disk is always larger than the amount of inertia of a sphere:

$$I(\text{sphere}) = \tfrac{2}{5}Ma^2 \qquad I(\text{disk}) = \tfrac{1}{2}Mb^2 \qquad (10.1)$$

where the mass of the system is constant as the gas collapses from a sphere with radius a to a disk of radius b. It is clear from this consideration that $b > a$, so that

$$I(\text{disk}) > I(\text{sphere}) \qquad (10.2)$$

for any process of this kind.

All of this describes a very complex process that can be simulated on a high-speed computer. This has actually been done, and the model described here when implemented on a computer leads to a planetary system similar to that which exists in the solar system. In fact, some of these calculations even predict the occurrence of terrestrial and giant planets that are observed in the solar system. The collapsing gas cloud model of the solar system's origin was first proposed by Immanuel Kant in the eighteenth century and worked out during the 1940s by C. F. Von Weizsäcker and G. P. Kuiper. It should be stressed, however, that the model described here can predict only qualitatively what happens and that the detailed placement of the planets by the mechanisms proposed here is a matter of chance and not one of deterministic and predictable physics.

One of the more amusing episodes in the history of the development of celestial mechanics is that in the late eighteenth century it was thought, for a time, that there was a deterministic mathematical way of establishing the distances of the planets from the Sun. Johannes Kepler tried to do this almost two hundred years earlier with his famous model of planetary orbits inscribed within the five "regular" polyhedra. This did not work in the end. However, during the 1770s, Johann Elert Bode and Johann Daniel Titius proposed an empirical "law" that gave the following relationship between the number of the planet and its distance from the Sun:

$$D_n = [(4 + 3 \times 2^n) \times 0.1] \times D_E \qquad (10.3)$$

where D_n stands for the distance of the nth planet from the Sun and D_E is the distance from Earth to the Sun. The distance D_E is usually called an *astronomical unit* and is roughly equal to 150×10^6 km. In order for the Bode–Titius law to work, some liberty must be taken with the definition of the exponent n in the sense that the value of n for Mercury must be set at minus infinity and for Venus at zero.

It should be mentioned that recently, Archie Roy has established a theoretical justification for the Bode–Titius law. The comparisons are shown in Table 10.2. The law is reasonably accurate out to the orbit of the planet Uranus. It misses the prediction for Neptune by 25% and for Pluto by a factor of 2. Most interesting of all is the fact that the Bode–Titius law predicts the existence of a planet at approximately the position where the asteroids are now found. The best guess today is that there was once such a planet but that it disintegrated because of the gravitational influence of Jupiter.

The 25 years from 1965 to 1990 can be called a "golden age" of planetary exploration. It was during this period that NASA executed a bold and

TABLE 10.2. Distance in Astronomical Units

Planet	n	Distance from Sun	Bode–Titius Prediction
Mercury	$-\infty$	0.3871	0.4
Venus	0	0.7233	0.7
Earth	1	1.0	1.0
Mars	2	1.5237	1.6
Asteroids	3	~3.0	2.8
Jupiter	4	5.2026	5.2
Saturn	5	9.5549	10.0
Uranus	6	19.2184	19.6
Neptune	7	30.1104	38.8
Pluto	8	39.5447	77.2

imaginative strategy to explore the entire solar system using robotic spacecraft. The strategy was clearly laid out in three phases:

1. Spacecraft would be sent to fly past each planet in the solar system. By 1990, this part of the strategy was almost completed because robot spacecraft had visited every planet except Pluto. These close fly-bys have yielded a wealth of important scientific data, and, perhaps equally important, they provided the data used for the planning of more sophisticated missions.

2. Spacecraft would be sent to orbit selected planets of special interest. Orbiters have been sent around Venus, Mars, and Jupiter. Complete maps of the planets Venus and Mars have been made as a result of these missions. An orbiter has recently (1995) been placed around Jupiter with the object of producing good maps of the four Galilean satellites.

3. Spacecraft would probe the atmospheres of selected planets and would also land on the surface of the planets whenever possible. Atmospheric probes have successfully examined the atmospheres of Venus, Mars, and Jupiter. Successful landings have been made on Mars and Venus, and in each case, pictures of the surface have been obtained.

The golden age of planetary exploration coincided with the height of the "cold war" between the United States and the Soviet Union. Therefore, a spirited competition in planetary exploration developed between the United States and the Soviet Union. The Soviet Union elected to concentrate efforts on Venus and Mars, leaving exploration of the outer solar system to the United States. The Soviet Venus exploration effort was extremely successful with both the first atmospheric entry into another planet and the first landing. On the other hand, their Mars program was essentially a failure. The United States adopted a more comprehensive strategy that called for an eventual visit to every planet in the solar system.

To execute these missions, four classes of spacecraft have been developed by the United States: the Mariners, the Pioneers, the Voyagers, and the Vikings. The first generation of the Mariner spacecraft was designed to explore the inner planets, Mars, Venus, and Mercury. *Mariner 2* flew past Mars in 1962 and returned the first high-resolution photo of another planet. Perhaps the most interesting Mariner mission conducted by the first generation of these spacecraft was the fly-by of Venus and Mercury by *Mariner 10* in 1973, in which a *gravity-assist* trajectory was used for the first time to permit the spacecraft to fly past both Venus and Mercury. A picture of the *Mariner 10* spacecraft that performed this mission is shown in Figure 10.3. All of the Mariners were three-axis, stabilized platforms using solar panels to provide the electric power. The scientific in-

FIGURE 10.3 *Mariner 10* spacecraft is shown in this picture. *Mariner 10* flew past both Venus and Mercury in 1974 and 1975 using a gravity-assist trajectory. *Mariner 10* is an example of a three-axis stabilized spacecraft used for the exploration of the inner planets.

strument packages consisted of a camera and various magnetic field detectors and charged particle counters. A picture of the planet Mercury made of a composite of the *Mariner 10* fly-by pictures is shown in Figure 10.4. *Mariner 10* made three fly-bys of Mercury, in March 1974, September 1974, and March 1975.

The second generation of Mariner spacecraft was more sophisticated. They were heavier (over 1000 kgm vs. 200–400 kgm) and were also three-axis stabilized. By far, the most important mission was by *Mariner 9,* the first spacecraft ever put in orbit around another planet. *Mariner 9* took thousands of high-resolution pictures of the Martian surface, which provided the information necessary to select the landing site for Viking.

The Viking spacecraft was definitely the most sophisticated produced up to 1975. It consisted of an orbiter and a lander that were put on a trajectory to Mars as a single unit. This spacecraft was placed in an orbit around Mars. At the appropriate time, the lander was separated from the spacecraft and then descended to the surface of the planet Mars. The orbiter continued to remain in Mars orbit to take pictures to augment and extend what was done with *Mariner 9.* The Viking landers were placed on the planet in 1976, and they were the first to make "in situ" analyses of the soil and the atmosphere of another planet. The Viking landers

FIGURE 10.4 Composite picture of Mercury taken by *Mariner 10* on March 29, 1974. The spacecraft approached to within 431 miles of the planet's surface. (Courtesy of the Planetary Society.)

also took spectacular pictures on the surface of Mars. A sample is shown in Figure 10.5, and a picture of the Viking lander is shown in Figure 10.6. An important part of the Viking lander mission was to look for biological activity on the surface of the planet. No evidence of biological activity was found. Figure 10.7 shows a composite picture of a Mercator projection of the surface of Mars taken by *Mariner 9* and the Viking orbiters.

In the field of planetary exploration during the 1970s and the 1980s, the Soviet Union concentrated on the planet Venus. About two dozen Venera spacecraft were launched by the Soviet Union starting in 1961. The most spectacular was *Venera 9,* which carried a probe that reached the surface of the planet in June 1975. A picture of the surface of Venus taken by *Venera 9* is shown in Figure 10.8. *Venera 9* was the first spacecraft to return a picture taken on the surface of another planet.

The United States also carried out an active program to explore the planet Venus. Pioneer Venus consisted of two spacecraft, an orbiter and a "multiprobe" spacecraft. A picture of the "multiprobe" spacecraft is shown in Figure 10.9. The multiprobe spacecraft had a spin-stabilized bus that carried four probes that were released simultaneously into the Venus atmosphere to measure wind patterns by determining the differential mo-

FIGURE 10.5 Picture of the Martian surface as taken by the *Viking 2* lander. This view looks out over Utopia Plain on the northern hemisphere of Mars. It is probably a typical example of Martian flatlands. (Courtesy of NASA.)

FIGURE 10.6 Viking Lander. This was one of the most sophisticated spacecraft ever built. The instrumentation on board included imaging, soil analysis, meteorological, and seismological equipment. In addition, there was an experiment designed to look for evidence of biological activity on Mars. (Courtesy of NASA.)

tion of the probes. A picture of Venus showing the dense cloud cover and the pattern of atmospheric motion is shown in Figure 10.10. (This picture was taken by Pioneer Venus in 1980.) High surface winds with speeds well over 161 km/h were measured by the multiprobe.

The Pioneer Venus orbiter carried a radar altimeter, which, along with a precise determination of the orbital ephemeris, led to the creation of the first good map of the planet Venus. The orbiter, like the multiprobe bus, was a small, relatively simple spin-stabilized spacecraft. The radar was necessary because, unlike visible light, radar can penetrate the thick cloud cover of the planet Venus. Figure 10.11 shows a globe of Venus constructed using the radar altimeter compared with what a globe of Earth would look like using the same instrumentation. The most significant finding of the Pioneer Venus orbiter was to confirm that the surface of Venus shows features that are very similar to those on Earth. There are

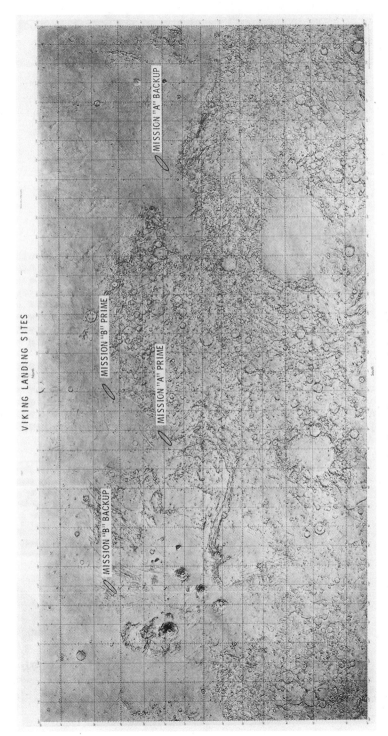

FIGURE 10.7 Mercator projection of a map of the planet Mars. It was produced from pictures taken by the Viking orbiters. (Courtesy of NASA.)

FIGURE 10.8 Picture of the surface of Venus taken with a camera mounted on *Venera 13.* (Courtesy of NASA.)

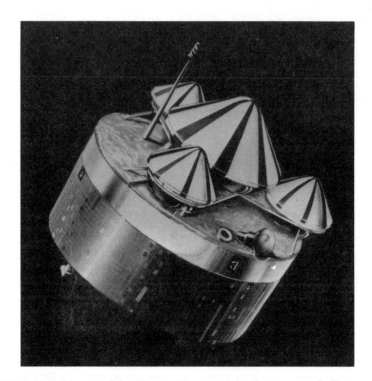

FIGURE 10.9 Picture of the Pioneer Venus multiprobe spacecraft. Four atmospheric entry probes are shown mounted on the spacecraft. The spacecraft is spin stabilized, and this allowed almost simultaneous deployment of the probes before the spacecraft reached Venus. All four probes survived atmospheric entry and returned valuable information about the dynamics of the Venus atmosphere. One of the probes survived the landing and continued to broadcast for about an hour.

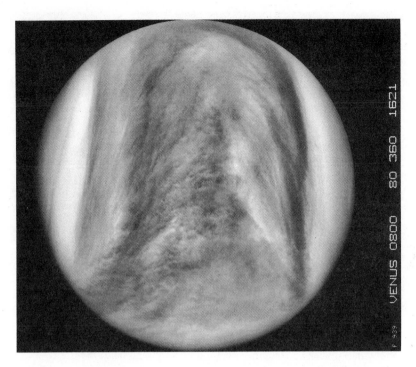

FIGURE 10.10 Picture of Venus taken by the Pioneer Venus Orbiter spacecraft. It is not possible to see the surface of the planet using visible light. Shows weather patterns in the Venus atmosphere typical of the upper layers of the atmosphere. (Courtesy of NASA.)

continents, mountain ranges, and rift valleys, and there is evidence that there are volcanic mountains.

The final chapter of the exploration of Venus was written by the *Magellan* spacecraft, which was launched in May 1989. *Magellan* was a large three-axis stabilized spacecraft, which was essentially an advanced Mariner design. The *Magellan* carried a high-resolution imaging radar that was able to take pictures of features on the surface with dimensions of the order of 10 m, which is 100 times better than the radar altimeter on the Pioneer Venus orbiter. Figure 10.12 shows a computer reconstruction of images taken by *Magellan* that illustrates what the surface of Venus would look like from a low-flying airplane. This spectacular picture is an excellent demonstration of what can actually be done with robotic spacecraft.

Planning for the exploration of the outer planets of the solar system

FIGURE 10.11 A globe of the planet Venus made from the radar altimeter data taken by the Pioneer Venus Orbiter compared with a globe of the Earth if it were made by a similar instrument. (Courtesy of NASA.)

(Jupiter, Saturn, Uranus, Neptune, and Pluto) began in the mid-1960s when it was recognized that in the late 1970s and early 1980s, the configuration of the outer planets would be peculiarly favorable, so that a single spacecraft might, with the use of gravity-assist procedures, fly past all of the outer planets. Thus, the concept of the "Grand Tour" was born. As things turned out, schedule delays made it impossible to execute the Grand Tour. However, by 1989, *Voyager 2*, which was essentially the follow-up to the Grand Tour, had flown past the planet Neptune, leaving Pluto the only planet in the solar system not visited by a robotic spacecraft. The Grand Tour using a Saturn launch vehicle was eventually abandoned. As a substitute, the Voyager program was initiated, which called for a very sophisticated spacecraft with three-axis stabilization, high-resolution cameras, and an array of other instruments intended to measure charged particle fluxes, plasmas, and magnetic fields.

Since the Voyager spacecraft would be very expensive—ultimately, the Voyager program would cost almost two billion dollars—it was decided in 1969 to initiate another outer planet exploration program called Pio-

FIGURE 10.12 Picture showing a computerized image of what the surface of Venus would look like from a high-flying aircraft. The picture was made using data taken by the *Magellan* spacecraft. (Courtesy of NASA.)

neer. The intent was to send two small and inexpensive spacecraft to the outer planets to answer two questions:

1. Could a spacecraft survive passing through the asteroid belt? In the 1960s, our knowledge of the mass distribution of particles in the asteroid belt was nonexistent, and it would be necessary to find out whether it would be safe to pass through.

2. What is the distribution and the flux of energetic charged particles trapped in Jupiter's magnetic field? Are the radiation levels high enough so that the spacecraft would be damaged or destroyed? It was known that Jupiter had a strong magnetic field, but the precise radiation levels created by the charged particles trapped in the field were not known.

The Pioneer spacecraft were designed to answer these questions. The Pioneer Outer Planet program was approved early in 1969, and it called for the development of two small spacecraft (250 kgm) that would be spin stabilized and that could be launched to reach Jupiter by a relatively, inexpensive launch vehicle, the Atlas-Centaur. The total program cost for Pioneer was capped at $100 million for two spacecraft to be called *Pioneer 10* and *Pioneer 11*.

Pioneer 10 was launched on March 3, 1972, and flew past Jupiter in December 1974. Because the spacecraft was on a gravity-assist trajectory, during which the velocity of the spacecraft was increased beyond the escape velocity out of the solar system in the encounter with Jupiter, *Pioneer 10* became the first man-made object to leave the solar system.

Figure 10.13 shows a picture of the *Pioneer 10* spacecraft. It is substantially different from the Mariners that were designed to look at the inner planets. The dominant feature of *Pioneer 10* and any outer planet spacecraft is the high-gain antenna necessary to send signals back from the vast distances that the outer planet spacecraft must reach. The second feature is that there are no solar cell panels on these spacecraft. The Sun is simply too far away beyond the orbit of Mars to power the spacecraft, and therefore nuclear power supplies, called radioisotope thermal generators (RTGs), must be used. These power supplies use the radioactive isotope, plutonium-238, as a heat source, and they can last a long time. The second Pioneer, *Pioneer 11,* was launched on April 6, 1973, and flew past Jupiter in December 1975. It was then put on a trajectory to fly past Saturn and became the first spacecraft to reach Saturn in September 1979 and to take close-up pictures of the planet's spectacular ring system.

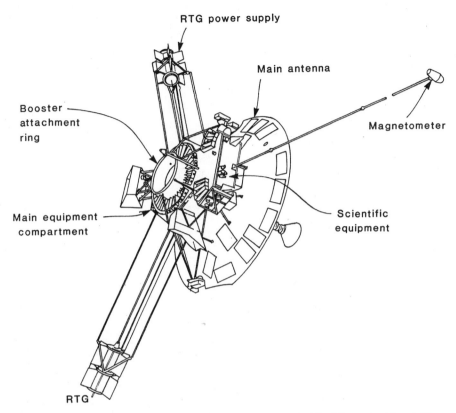

RTG power supply

Main antenna

Booster
attachment
ring

Magnetometer

Main equipment
compartment

Scientific
equipment

RTG

FIGURE 10.13 Picture shows a line drawing of the identical *Pioneer 10* and *Pioneer 11* spacecraft. *Pioneer 10* was the first spacecraft to fly past the orbit of Mars, fly through the asteroid belt, conduct a close fly-by of Jupiter, and then be put on a trajectory to leave the solar system. *Pioneer 11* was the first spacecraft to conduct close fly-bys of both Jupiter and Saturn. The Pioneer spacecraft were spin stabilized and used a radioisotope-fueled (plutonium-238) power supply. *Pioneer 10* is now (1996) a little less than 5 billion miles (50 a.u.) from Earth and is still in contact with the ground controllers. (From *Pioneer Odyssey,* by R. O. Fimmel, W. Swindell, and E. Burgess, NASA SP-396, 1977.)

In addition to the scientific results that were obtained, the Pioneer spacecraft also answered the questions that are listed on page 204. It was demonstrated both that it is safe to fly a spacecraft through the asteroid belt and that a properly designed spacecraft can fly past Jupiter without suffering unacceptable damage. Thus, the Voyager program was given the go-ahead. As in the case of Pioneer, there were two spacecraft, *Voyager 1* and *Voyager 2. Voyager 1* was launched on September 5,

1977, and it produced some of the most spectacular pictures yet taken of Jupiter and Saturn and their satellites. *Voyager 2* executed the most comprehensive space voyage ever performed by an exploratory vehicle. Launched on August 20, 1977, the spacecraft flew past Jupiter (on July 9, 1979), Saturn (on August 25, 1981), Uranus (on January 4, 1986), and Neptune (on August 15, 1989). Voyager 2 is now also on a trajectory that will take it out of the solar system. Two samples of the kind of results that were obtained are shown in Figure 10.12 and 10.13. In Figure 10.14, we show a volcanic eruption on the surface of Jupiter's satellite, Io. Figure 10.15 shows a spectacular picture of the Saturn ring system that reveals its enormously complex structure. *Voyager 2* is definitely one of the most successful space exploration programs ever executed.

We are now ready to do some calculations. The purpose of the previous sections has been to describe the exploration of the solar system that

FIGURE 10.14 Picture shows a volcanic eruption on the Jovian satellite, Io. The plume extends 100 miles above the surface of the satellite. (Courtesy of NASA.)

FIGURE 10.15 This picture of Saturn's ring system illustrates its great complexity. It was taken by *Voyager 2*. (Courtesy of NASA.)

has been performed in general terms. Some of the maneuvers that spacecraft must execute to perform these missions will be described in some detail. We shall first determine the "escape" velocity from the solar system starting from Earth. Second, we will examine the trajectories of spacecraft as they fly from one planet to another and particularly the time that it takes to make interplanetary trips. Finally, we will examine how spacecraft can pick up or lose speed in encounters with planets moving in their orbits. This is the so-called slingshot effect.

Let us start by looking at some numbers. The first problem is to estimate the influence of the Sun. In dealing with Earth-orbiting spacecraft, we have the following number:

$$v_E = \sqrt{\frac{GM_E}{R_E}} = 7905 \text{ km/s} = 28{,}458 \text{ km/h} \tag{10.4}$$

where M_E is the mass of Earth and R_E is the radius of Earth. This is a known number, and it is the velocity that an object must have in order to sustain a circular orbit if the object could move at the radius R_E. The "gravitational influence" of Earth might be defined as

$$\sqrt{GM_E} = 28,458 \sqrt{R_E} \tag{10.5}$$

and $R_E = 6371$ km, so that we can write for this constant

$$\sqrt{GM_E} = 28,458 \times 79.8 \approx 2.27 \times 10^6 \tag{10.6}$$

where this constant is measured in the "metric" units that we have been using.

We can use the motion of Earth around the Sun to calculate the same constant for the Sun, or what we might call the gravitational influence of the Sun. In this case,

$$v_0 = \sqrt{\frac{GM_s}{R_0}} \tag{10.7}$$

where v_0 is the velocity of Earth in orbit, R_0 the radius of Earth's orbit around the Sun, and M_S the mass of the Sun. The assumption in equation (10.7) is that the orbit of Earth is approximately circular. The numbers can now be evaluated as follows:

$$\sqrt{GM_s} = v_0\sqrt{R_0} \tag{10.8}$$

and we have also, for the period of Earth's orbit, T_0, around the Sun,

$$\frac{2\pi R_0}{v_0} = T_0 = 1 \text{ year} \tag{10.9}$$

Now the radius of Earth's orbit is

$$R_0 = 1.496 \times 10^8 \text{ km} \tag{10.10}$$

and the period is

$$T_0 = 1 \text{ year} = 365 \times 24 = 8.76 \times 10^3 \text{ h} \tag{10.11}$$

Thus we have

$$v_0 = \frac{2\pi R_0}{T_0}$$

$$= 107{,}200 \text{ km/h} \tag{10.12}$$

which is almost four times larger than the orbital velocity of a spacecraft moving around Earth in a low altitude orbit.

The gravitational influence of the Sun can now be estimated in the same way that was done for the case of Earth in equation (10.5):

$$\sqrt{GM_S} = v_0\sqrt{R_0}$$

$$= 107{,}200 \times 12.2 \times 10^3 = 1.31 \times 10^9 \tag{10.13}$$

Thus the gravitational influence of the Sun is about 577 times as great as that of Earth.

From equations (10.6) and (10.13), it is possible to estimate the ratio of the Sun's mass to that of Earth:

$$M_S = (577)^2 M_E$$

$$= (333{,}000)M_E \tag{10.14}$$

This is approximately the same as the correct ratio.

The escape velocity of a spacecraft can now be estimated, where by "escape" we mean escape from the solar system. This calculation must be performed assuming that the spacecraft already has Earth orbital velocity of 107,200 km/h. What we need to do then is to calculate the velocity that the spacecraft needs to acquire to escape completely from the solar system. To do that, we will refer to Figure 10.16. Earth's orbit will be assumed to be approximately circular with a radius R_0 of about 149.6 × 10^6 km. We will assume that the spacecraft is in a near-Earth equatorial orbit around Earth. The question is then what additional velocity increment must the spacecraft acquire in order to get into an escape orbit from the Sun. From Figure 10.16, it is obvious that the total velocity of the spacecraft at the point in the spacecraft orbit where the two velocities add is

$$v(\text{spacecraft}) = 107{,}266 + 27{,}750 = 135{,}016 \text{ km/h} \tag{10.15}$$

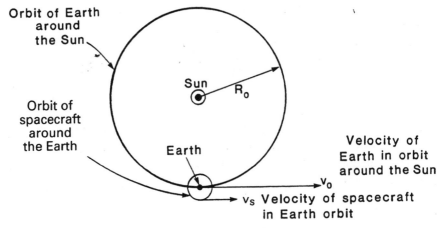

FIGURE 10.16

To escape from the solar system, the lowest *energy* orbit that permits escape is a *parabolic* orbit, which would have a closest distance of approach to the Sun equal to the radius of Earth. We must now determine the velocity of an object in such an orbit, and then the velocity increment required is the difference between that velocity and the velocity that the spacecraft already has as a result of its own motion around Earth and the motion of Earth around the Sun [see equation (10.15)].

The total energy of a parabolic orbit is zero, as we have shown in Chapter 6, so that we can equate the potential and kinetic energy at any point in the orbit:

$$\tfrac{1}{2}mv^2(\text{escape}) = \frac{GM_s m}{R_0}$$

and so

$$v(\text{escape}) = \sqrt{2}\sqrt{\frac{GM_s m}{R_0}} \qquad (10.16)$$

Therefore the velocity increment required for escape is

$$\Delta v = v(\text{escape}) - (v_0 + v_E)$$

$$= 151{,}674 - 135{,}016 = 16{,}658 \text{ km/h}$$

This is a substantial velocity increment that is necessary to get out of the solar system. It is *not easy* to get away from the Sun!

The final topic that we wish to discuss has to do with the so-called slingshot effect. This is the phenomenon that occurs when a spacecraft encounters a planet and gains or loses velocity that permits it to perform the next portion of the mission. The slingshot effect was first used by the *Pioneer 10* spacecraft during the Jupiter encounter on December 4, 1973, to permit the spacecraft to achieve a velocity high enough to get out of the solar system. Subsequently on February 5, 1974 *Mariner 10* also used the slingshot effect to get from the orbit of Venus to the orbit of Mercury. In this case also, the trajectory of the spacecraft had to be adjusted to gain velocity. The detailed calculation is complex because of the nature of the orbit, but the principle can be described quite easily. In the case of *Pioneer 10,* the problem was to fly by Jupiter to gain velocity. The velocity gain comes about because of the conservation of momentum. Since the planet is much larger than the spacecraft, its recoil can be essentially ignored. We will first look at a very specific case of the spacecraft's encounter with Jupiter to explain the principle.

We start by looking at Figure 10.17. The first step is to put the space-

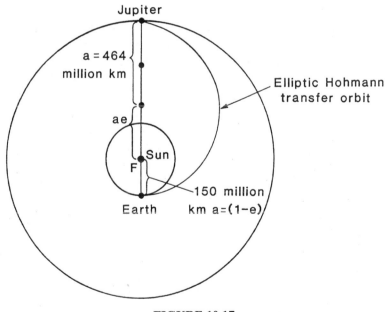

FIGURE 10.17

craft into a Hohmann transfer orbit as shown. The orbital velocity of Jupiter, v_J, can be calculated from the data on Jupiter in Table 10.1:

$$v_J = \sqrt{\frac{GM_s}{R_J}} \tag{10.17}$$

Now, the radius R_J of the orbit of Jupiter is given as

$$R_J = 5.2 \times 149{,}637$$

$$= 778{,}000 \times 10^6 \text{ km} \tag{10.18}$$

The velocity of Jupiter in orbit is therefore

$$v_J = \frac{1.31 \times 10^9}{\sqrt{778 \times 10^6}} = \frac{1.31}{27.9} \times 10^6$$

$$= 46{,}953 \text{ km/h} \tag{10.19}$$

As expected, the velocity of Jupiter is smaller than the velocity of Earth in its orbit.

As we saw in Chapter 8, the velocity of the spacecraft when it reaches Jupiter in the Hohmann transfer orbit is *smaller* than the velocity of an object in a *circular* orbit at the radius corresponding to apogee. From equation (8.29), we have, by a simple manipulation to transform from perigee to apogee,

$$v_T = \sqrt{\frac{GM_s}{R_J}} \sqrt{\frac{1-e}{1+e}} \tag{10.20}$$

where e is the eccentricity of the Hohmann transfer orbit. Now, the eccentricity of the transfer orbit can be worked out from Figure 8.4. The semi-major axis of the ellipse corresponding to the Hohmann transfer orbit is

$$2a(\text{Jupiter transfer}) = 778 + 150 = 928 \times 10^6 \text{ km}$$

$$= 464 \times 10^6 \text{ km} \tag{10.21}$$

From Figure 10.17, it can be seen that we can calculate the eccentricity as follows:

$$a(\text{Jupiter transfer})e + R_0 = a(\text{Jupiter transfer}) \tag{10.22}$$

and substituting the numbers yields

$$1 - e = \frac{R_0}{a(\text{Jupiter transfer})} = \frac{150}{464}$$

$$e = 1 - \frac{150}{464} = 1 - 0.323 = 0.677 \qquad (10.23)$$

Thus, the velocity at apogee is

$$v_A(\text{transfer orbit}) = v_J \sqrt{\frac{1-e}{1+e}} = 46{,}953 \sqrt{\frac{0.323}{1.677}}$$

$$= 20{,}612 \text{ km/h} \qquad (10.24)$$

The velocity at apogee is therefore substantially smaller than the velocity of Jupiter in its orbit.

Before we look at the slingshot effect, there are two other points about the trip to Jupiter that need to be considered. One is how often the opportunity occurs to go to Jupiter. The answer is approximately once a year since the period of Jupiter's orbit is 11.9 years. Thus, Earth moves much more rapidly in its orbit than Jupiter so that for all practical purposes Jupiter stands still. Actually, it turns out that the opportunity arises once every 13 months, as we can see from Figure 10.18.

The second question that we should answer is how long it takes the spacecraft to get from Earth to Jupiter. The period of the elliptic Hohmann transfer orbit can be calculated from Kepler's law using the fact that the period depends only on the semimajor axis. Thus, we can compare the period of the transfer orbit to the period of Earth in its orbit, T_0, using the following relationship:

$$\frac{T^2(\text{transfer orbit})}{T_0^2} = \frac{a^3(\text{transfer orbit})}{R_0^3} \qquad (10.25)$$

which reduces to

$$T(\text{transfer orbit}) = T_0 \left[\frac{(464)^3}{(150)^3} \right]^{1/2}$$

$$= 5.44 \text{ years} \qquad (10.26)$$

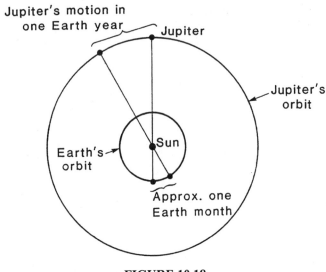

FIGURE 10.18

Now, since the spacecraft executes only *half* the transfer orbit, the time it takes to get to Jupiter is

$$T = 2.72 \text{ years} \qquad (10.27)$$

The last point now is to deal with the slingshot effect. This is quite complicated since the gravitational field of Jupiter complicates the situation, as shown in Figure 10.19. Jupiter is moving faster than the space-

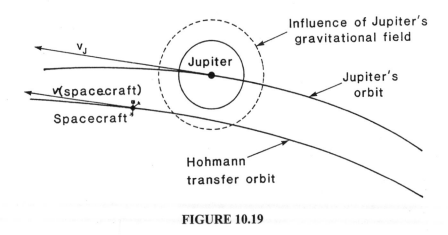

FIGURE 10.19

craft [see equations (10.19) and (10.24)], so that the problem is to place the spacecraft in front of the planet and to let the planet catch up. The mechanism is then to let Jupiter transfer momentum to the spacecraft. The actual rigorous calculation of this effect is rather complex since it will involve the detailed calculation of the orbit of the spacecraft around the planet Jupiter. We will simplify the situation by assuming that Jupiter is a hard sphere and that the momentum transfer calculation can be made by assuming an elastic collision with the hard sphere. We will also assume that the gravitational field within the planet's "influence" will pull the spacecraft toward the planet. This velocity component is not important in the case when the spacecraft must be speeded up, but we will see that it is necessary in the case when we want to slow the spacecraft down.

Figure 10.20 illustrates the situation described in the previous paragraph. It is obvious that the velocity of the spacecraft relative to Jupiter is such that the spacecraft is moving toward Jupiter. Let us transform to a coordinate system in which Jupiter is at rest. The relative velocity of the spacecraft is

$$v(\text{spacecraft-relative initial}) = -v_J + v(\text{spacecraft}) \qquad (10.28)$$

Now, if the spacecraft makes an elastic collision with Jupiter, the velocity is reversed so that the final velocity of the spacecraft after the collision is

$$v(\text{spacecraft-relative final}) = -[-v(\text{Jupiter} + v(\text{spacecraft})] \qquad (10.29)$$

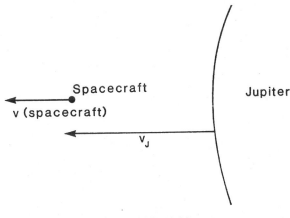

FIGURE 10.20

Now, transforming back to the coordinate system in which Jupiter is moving, we have, for the final velocity of the spacecraft in the original coordinate system,

$$v(\text{spacecraft-final}) = 2v_J - v(\text{spacecraft}) \qquad (10.30)$$

During the collision, therefore, the spacecraft picks up twice the velocity of Jupiter. Thus, the final velocity of the spacecraft is

$$v(\text{spacecraft final}) = 93{,}906 - 20{,}612 = 73{,}294 \text{ km/h} \qquad (10.31)$$

The question is whether this velocity is large enough to escape from the solar system. What we need to do is to calculate the escape velocity from the orbit of Jupiter. Equation (10.16) says that

$$v(\text{escape}) = \sqrt{2}\sqrt{\frac{GM_s}{R_{0J}}}$$

where R_{0J} is the radius of Jupiter's orbit. Thus

$$v(\text{escape}) = 66{,}392 \text{ km/h} \qquad (10.32)$$

From this calculation, it can be seen that the velocity of the spacecraft after the collision exceeds the escape velocity. This is how *Pioneer 10* became the first man-made object to leave the solar system. The actual trajectories of *Pioneer 10* and *Pioneer 11* are shown in Figure 10.21. The gravity-assist trajectories of both spacecraft are clearly shown as they pass by each planet. Both spacecraft passed the orbit of Pluto in 1990. *Pioneer 10* is now a little less than five billion miles from Earth (about 50 a.u.), and until March 31, 1997, it was in contact with the spacecraft's controllers at the NASA-Ames Research Center. Contact with the spacecraft was terminated not because the spacecraft failed but because of higher priority missions requiring more of the limited resources of the deep space communication network.

A similar calculation can be performed to show how a spacecraft can lose speed in a collision. In this case, the spacecraft approaches the planet from the right rather than the left. Thus, the influence of gravity is important because without the velocity increment provided by gravity the spacecraft will never reach the planet. It is, of course, this point that makes the calculation of the velocity decrement more complex than that for the velocity increment.

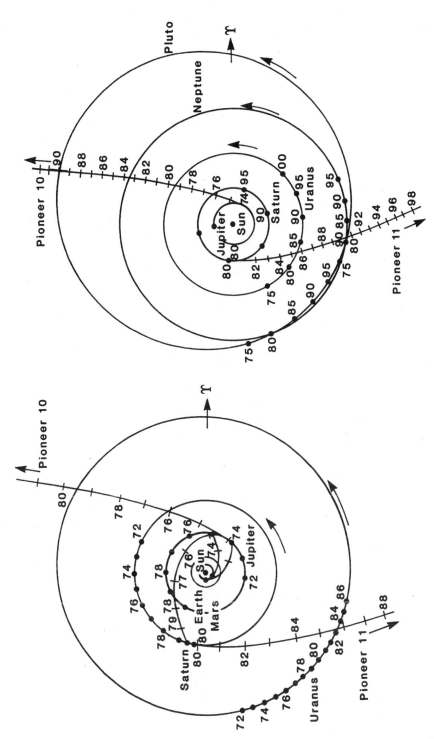

FIGURE 10.21 Drawing of the gravity-assist trajectories of *Pioneer 10* and *Pioneer 11* in their flights past Jupiter and Saturn. Both spacecraft passed beyond the orbit of Pluto in 1990.

217

By now many good works on planetary exploration are available that would be appropriate for further reading. At the risk of being somewhat arbitrary, the following books are recommended. A classic in the field is Nobel Laureate Harold Urey's *The Planets: Their Origin and Development* published in 1952. This book is an excellent description of the state of knowledge before the era of planetary exploration with space probes began. A good place to start learning about the detailed technical results of some of the early planetary spacecraft flights is in the *Evolution of the Solar System* edited by Hannes Alfven and Gustav Arrhenius. In addition, separate books on each planet, that has been investigated in more detail, are also available.

Mars is, of course, a special case because we have sent three orbiters (*Mariner 9* in 1971 and *Viking Orbiters I* and *II* in 1976) and have landed robotic spacecraft on the surface of the planet three times (*Vikings I* and *II* in 1976 and *Mars Pathfinder* in 1997). Two comprehensive works are *On Mars: Exploration of the Red Planet 1958–1978* by Edward C. Ezell and Linda N. Ezell and "Scientific Results of the Viking Project" published by the American Geophysical Union in 1977 which is a collection of papers on the subject of Viking that appeared in *The Journal of Geophysical Research* in 1977. There are some excellent compilations of pictures taken of the Martian surface by *Mariner 9* and the *Viking Landers*. The best of these are *The Martian Landscape* by the *Viking Lander* Imaging Team and *The Channels of Mars* by Victor R. Baker. In addition to the technical literature, there are good popular works about Mars that make worthwhile reading. These include *Mars Beckons* by John Noble Wilford and *The Planet Mars: A History of Observation and Discovery* by William Sheehan. No list of books about Mars would be complete without mentioning Percival Lowell's *Mars and Its Canals* published in 1906 in which the author describes his observations and then extrapolates in an imaginative but unfortunately unwarranted manner to suggest that Mars is inhabited by intelligent beings.

The literature on Venus is also extensive but less oriented toward the general public because Venus is not as exciting as Mars in terms of ultimately sending people to the planet. A good early description of the exploration of Venus using spacecraft is *The Venus Atmosphere* edited by Robert Jastrow and S. I. Rasool published in 1969. Subsequent results of the *Pioneer Venus* flights are described in "Pioneer Venus," a collection of papers from *The Journal of Geophysical Research* published in 1980 and *Pioneer Venus* by Richard O. Fimmel, Lawrence Colin and Eric Burgess. The *Magellan* results are presented in *Venus: The Geological Story* by Peter Cattermole, The Johns Hopkins University Press (Baltimore), 1994.

With respect to the outer planets, there are a number of books that are worth examining. A comprehensive early work is *Jupiter* edited by Tom Gehrels. In addition, NASA has published some excellent books on the *Pioneer* and the *Voyager* missions to the outer planets. These include, *Pioneer Odyssey* by Richard O. Fimmel, William Swindell and Eric ·Burgess; *Pioneer: First to Jupiter, Saturn, and Beyond* by Richard O. Fimmel, James A. Van Allen and Eric Burgess; *Voyage to Jupiter* (1980) and *Voyage to Saturn* (1982) both by David Morrison. A complete collection of papers published in *The Journal of Geophysical Research* as a result of the *Pioneer 11* fly-by of Saturn was published in 1980 by the American Geophysical Union. An excellent recent work on Jupiter is *The Giant Planet Jupiter* published in 1995 by John H. Rogers. Gary Hunt and Patrick Moore published an excellent illustrated study of the ringed-planet called *Saturn* in 1982. There are several good books on Uranus and Neptune. Among these are *Uranus: The Planet, Rings and Satellites* by Ellis D. Miner published in 1990; *The Planet Neptune* by Patrick Moore published in 1988; and *Uranus and Neptune: The Distant Giants* (1988) and *Far Encounter: The Neptune System* (1991) both by Eric Burgess.

PROBLEMS

10.1. The semimajor axis of the orbit of Neptune is 30 a.u. and that of Pluto is 39.5 a.u. The eccentricity of Neptune's orbit is 0.01 and that of Pluto is 0.25.

(a) Show that a portion of Pluto's orbit is located inside the orbit of Neptune.

(b) What fraction of Pluto's orbital period is spent inside the orbit of Neptune? (Assume that the orbit of Neptune is circular for the purpose of this calculation.)

10.2. The influence of the Sun dominates the motion of the planets in the solar system. However, the planets also influence each other through the gravitational forces that they exert on each other. The maximum force that a planet can exert is when the planet is in the position shown in the accompanying diagram. The closest approach is the distance $R_P - R_E$. Earth is most heavily influenced by four planets, Venus, Mars, Jupiter, and Saturn.

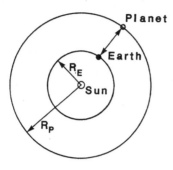

Planet

Closest approach
is the distance
(R_P-R_E)

(a) Which of these planets exerts the largest maximum force on Earth?

(b) Which of the planets exerts the smallest maximum force on Earth?

10.3. An asteroid that has one-tenth of the mass of the Moon strikes the Moon, as shown in the accompanying diagram. The moon is moving with a velocity v_M with respect to Earth, and the asteroid is moving toward the Moon along the line joining the centers of the two bodies with a velocity v_A, which is equal in magnitude to the velocity of the Moon but in the opposite direction. In the collision, the entire momentum of the asteroid is transferred to the Moon. (Another way of saying this is that the collision is completely "inelastic.") Assume that initially the Moon is a circular orbit around the Earth with a radius of about 384,000 km.

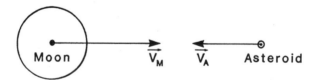

(a) What is the velocity of the Moon before and after the impact of the asteroid on the Moon?

(b) After the collision, the Moon will be in an elliptic orbit around Earth. Determine the semimajor axis of the new orbit and also the eccentricity.

CHAPTER 11

GENERAL PERTURBATION THEORY AND A SPECIFIC APPLICATION TO THE MOTION OF THE PLANET MERCURY

In prior chapters, we have always made the assumption that we are dealing only with two bodies interacting with each other through a gravitational field. In reality, this situation never occurs, and we have used the two-body approximation first to illustrate basic principles and then to develop and illustrate some of the methods used for the determination of orbits (Kepler's equation and Lambert's theorem). Now we have to look at real situations and to determine how best to include the effects that we have so far ignored.

In the next two chapters, we will consider two important cases, the solar system and the behavior of Earth-orbiting artificial satellites. In each of these problems, the approximations we have made in the past break down. The solar system consists of more than two bodies, and in developing more accurate orbits for the planets, it is necessary to consider not only the influence of the Sun but also that which the other planets orbiting the Sun exert on the planet whose orbit we are trying to determine. Not unexpectedly, we shall learn that the largest planet in the solar system, Jupiter, is the most important and that Jupiter exerts measurable influence on the orbits of the other planets in the solar system. This influence is still relatively small because the Sun contains 99.86% of the mass of the solar system and Jupiter only about 0.1%. Nevertheless, the accuracy of modern astronomical techniques makes it possible to make significant measurements of even very small deviations from the two-body results.

Somewhat similar considerations apply to Earth-orbiting satellites. In this case, Earth is the dominant source of the gravitational field that determines the behavior of the satellite. However, both the Moon and the Sun are large enough to exert enough influence on the motion of the satellite so that it can be measured. Once again, the fact is that we are not dealing with a pure two-body situation. An even more important effect results from the fact that Earth is not a perfect sphere, nor is it perfectly homogeneous. We showed in Chapter 2 that the source of Earth's gravitational field can be considered as located at the center of the sphere. Because of the centrifugal forces that result from Earth's rotation on its axis, Earth is an *oblate spheroid* rather than a sphere. This means that the diameter of Earth measured in the equatorial plane is somewhat larger than the diameter measured along the axis of rotation that joins the north and the south pole (the difference is about 21 km). Another important factor that affects the orbits of Earth satellites is that Earth's mass is not distributed homogeneously. There is, of course, a radial dependence of the mass as measured from Earth's center with the largest density at the center of Earth and decreasing as we move out from the center. If this were all, then there would be no effect since the mass of Earth could still be considered located at Earth's center. (This can easily be seen by looking at the derivation in Chapter 2.) However, Earth's mass distribution also depends on the latitude and the longitude, that is, on the azimuthal angle (θ) and on the polar angle (ϕ). This means that Earth's center of mass is no longer located at the geometric center of the oblate spheroid so that the behavior of satellite orbits is more complicated than the two-body result.

In the case of the nonspherical and nonhomogeneous Earth, we are again fortunate in the sense that the deviations from perfect, uniform sphericity are small. Once again, our ability to make very accurate determinations of satellite orbits makes it possible to use satellite orbital measurements to learn about the structure of Earth even though the deviations from orbits for a perfectly spherical Earth are small. We shall now describe mathematical techniques to calculate the orbits of planets and satellites when small deviations from the "perfect" two-body problem exist. These techniques are described by the general term *perturbation theory*.

The fundamental idea of perturbation theory can be described as follows: If a given problem can be approximated by a two-body problem, then it is possible to use the exact solution of the two-body problem to derive equations that have as a solution the difference between the two-body solution and that produced by the small perturbation. In this way, it is possible to determine the changes in the orbit that the perturbation produces. To illustrate this procedure, we shall start by looking at an example that makes the principle of perturbation theory transparent: the simple harmonic oscillator.

The simple harmonic oscillator is the one-dimensional mechanical system illustrated in Figure 11.1. It consists of a mass m attached to a spring with a spring constant k. When the spring is neither extended nor compressed, the mass m is located at the point $x = 0$. If the mass is set in motion and then released, the spring will exert a force on the mass that is proportional to the displacement of the mass from the equilibrium point $x = 0$. Throughout the motion, the force exerted by the spring will always be exactly equal and opposite in direction to the inertial force (the time rate of change of the momentum) of the mass. If there are no dissipative forces (frictional forces as the mass slides along the plane or energy dissipation as the spring extends and compresses), then the equation of motion is

$$m\frac{d^2x}{dt^2} = -kx \quad \text{or} \quad m\frac{d^2x}{dt^2} + kx = 0 \qquad (11.1)$$

The general solution of this equation is

$$x = a \sin \omega t + b \cos \omega t \qquad (11.2)$$

in which the amplitudes a and b are appropriate constants of integration. Equation (11.2) says that the mass oscillates around the point $x = 0$ with an amplitude A and a phase angle θ_0 defined by the equation

$$x = A \sin(\omega t + \theta_0) \qquad (11.3)$$

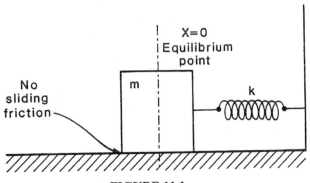

FIGURE 11.1

where the amplitude A and the phase angle θ_0 can be related to the constants of integration a and b:

$$a = A \cos \theta_0 \quad \text{and} \quad b = A \sin \theta_0 \tag{11.4}$$

The motion described by equation (11.3) is a simple oscillation of the mass with a frequency defined by ω. Substituting equation (11.3) into equation (11.1) yields the following expression for the frequency:

$$\omega = \sqrt{\frac{k}{m}} \tag{11.5}$$

The relationships we have developed are valid if there are no forces other than the spring acting on the oscillating mass. If we introduce a time-dependent external force $f(t)$, then the equation of motion is

$$m\frac{d^2x}{dt^2} + kx = f(t) \tag{11.6}$$

For the purpose of developing a useful perturbation theory, it is convenient to rewrite the equation of motion (11.6) as two coupled, first-order equations, rather than as one second-order equation. Let

$$y = \frac{dx}{dt} \tag{11.7}$$

then we have, by substituting (11.7) into (11.6),

$$m\frac{dy}{dt} + kx = f(t) \tag{11.8}$$

Equations (11.8) and (11.7) constitute the set of coupled equations we need. Using equation (11.5), we can rewrite equation (11.8) as

$$\frac{dy}{dt} + \omega^2x = \frac{f(t)}{m} = g(t) \tag{11.9}$$

where $g(t)$ is the time-dependent acceleration produced by the force $f(t)$. If the external force is zero, then the solutions of the coupled equations are, from equation (11.2),

$$x = a \sin \omega t + b \cos \omega t \tag{11.10}$$

and

$$y = a\omega \cos \omega t - b\omega \sin \omega t \qquad (11.11)$$

We are now ready to apply the principle of perturbation theory that we have stated previously. What we will assume is that when the time-dependent force is applied to the oscillator, then the form of the solution will not change much from that of equation (11.10). The solution will still look like this:

$$x = a(t)\sin \omega t + b(t)\cos \omega t \qquad (11.12)$$

but now, $a(t)$ and $b(t)$ are no longer "constants." They must be treated as functions of time. However, if the solution (11.12) is to retain the same form as the solution to equation (11.1), that is, when no force is present, then $a(t)$ and $b(t)$ must be *slowly varying* functions of time. If this condition is not fulfilled, then the assumption that the same form can be used for the solution as in the case when $f(t) = 0$ breaks down.

The final step is to derive differential equations for $a(t)$ and $b(t)$. To do that, we will substitute the solution (11.12) in equations (11.7) and (11.8) and retain the assumption that $x(t)$ and $y(t)$ retain their functional forms. Thus

$$y(t) = a(t)\omega \cos \omega t - b(t)\sin \omega t = \frac{dx}{dt} = \frac{da(t)}{dt}\sin \omega t$$

$$+ a(t)\omega \cos \omega t + \frac{db(t)}{dt}\cos \omega t - b(t)\omega \sin \omega t \qquad (11.13)$$

or

$$\frac{da(t)}{dt}\sin \omega t + \frac{db(t)}{dt}\cos \omega t = 0 \qquad (11.14)$$

Using equation (11.13), we can write, for the time derivative of $y(t)$,

$$\frac{dy(t)}{dt} = \frac{da(t)}{dt}\omega \cos \omega t - a(t)\omega^2 \sin \omega t$$

$$- \frac{db(t)}{dt}\omega \sin \omega t - b(t)\omega^2 \cos \omega t \qquad (11.15)$$

Using equations (11.15) and (11.12) in equation (11.9), we obtain

$$\frac{da(t)}{dt}\omega\cos\omega t - a(t)\omega^2\sin\omega t - \frac{db(t)}{dt}\omega\sin\omega t - b(t)\omega^2\cos\omega t$$

$$+ \omega^2 a(t)\sin\omega t + \omega^2 b(t)\cos\omega t = g(t) \qquad (11.16)$$

Simplifying yields

$$\frac{da(t)}{dt}\omega\cos\omega t - \frac{db(t)}{dt}\omega\sin\omega t = g(t) \qquad (11.17)$$

If we now multiply equation (11.16) by cos ωt and equation (11.15) by ω sin ωt and add the equations, we have

$$\omega\frac{da(t)}{dt}(\sin^2\omega t + \cos^2\omega t) = \omega\frac{da(t)}{dt} = g(t)\cos\omega t \qquad (11.18)$$

Likewise, multiplying (11.16) by sin ωt and (11.15) by ω cos ωt, we obtain

$$\omega\frac{db(t)}{dt} = -g(t)\sin\omega t \qquad (11.19)$$

Rewriting these yields two differential equations for the time-dependent coefficients $a(t)$ and $b(t)$:

$$\frac{da(t)}{dt} = \frac{g(t)}{\omega}\cos\omega t \qquad (11.20)$$

and

$$\frac{db(t)}{dt} = -\frac{g(t)}{\omega}\sin\omega t \qquad (11.21)$$

It is worthwhile to repeat here that implicit in this procedure is the assumption that the form of the solution remains the same when the external force is applied. This means that the external force must be small compared to the force exerted on the mass by the spring.

We are now ready to use this formulation to calculate what happens when different kinds of perturbing forces are applied. The simplest assumption is that the applied external force is constant, that is, not a function of time. In that case

$$f(t) = mg(t) = m\gamma \qquad (11.22)$$

where γ is the small acceleration produced by the external force. Since the perturbation is small, we will assume that $a(t)$ and $b(t)$ can be written as

$$a(t) = a + \alpha(t) \qquad (11.23)$$

and

$$b(t) = b + \beta(t) \qquad (11.24)$$

where a and b are the amplitudes of the unperturbed motion [that is, $f(t) = 0$], and the functions $\alpha(t)$ and $\beta(t)$, which describe the time dependence imposed by the external force, are always small compared to a and b. Using equations (11.23) and (11.24) in equations (11.19) and (11.20), we have

$$\frac{d\alpha(t)}{dt} = \frac{\gamma}{\omega}\cos \omega t \qquad (11.25)$$

and

$$\frac{d\beta(t)}{dt} = -\frac{\gamma}{\omega}\sin \omega t \qquad (11.26)$$

Integrating these equations yields, for $\alpha(t)$ and $\beta(t)$,

$$\alpha(t) = \frac{\gamma}{\omega^2} \int \cos \omega t \, d(\omega t) = \frac{\gamma}{\omega^2} \sin \omega t \qquad (11.27)$$

and

$$\beta(t) = \frac{\gamma}{\omega^2} \int -\sin \omega t \, d(\omega t) = \frac{\gamma}{\omega^2}\cos \omega t \qquad (11.28)$$

Therefore

$$a(t) = a + \frac{\gamma}{\omega^2}\sin \omega t \qquad (11.29)$$

and

$$b(t) = b + \frac{\gamma}{\omega^2} \cos \omega t \tag{11.30}$$

Finally, substituting (11.20) and (11.29) into the solution of the perturbed equation yields

$$x(t) = a \sin \omega t + \frac{\gamma}{\omega^2} \sin^2 \omega t + b \cos \omega t + \frac{\gamma}{\omega^2} \cos^2 \omega t$$

and therefore

$$x(t) = a \sin \omega t + b \cos \omega t + \frac{\gamma}{\omega^2} \tag{11.31}$$

The motion of the perturbed harmonic oscillator is therefore represented by the sum of two terms, the first term on the right side, which is the solution of the unperturbed oscillator, and another term that describes the change due to the perturbation. If $x_0(t)$ is the solution of the unperturbed oscillator, then we can write, for (11.30),

$$x(t) = x_0(t) + \frac{\gamma}{\omega^2} \tag{11.32}$$

or more generally,

$$x(t) = x_0(t) + \delta x(t) \tag{11.33}$$

where the second term on the right side, $\delta x(t)$, is the change induced by the perturbation. The function $x_0(t)$ is the unperturbed solution for the harmonic oscillator. Therefore, by following the procedures outlined, we have done what we have said about the form of perturbation solutions.

Let us return to the problem we have been considering, a perturbation represented by the addition of the small constant force $m\gamma$. We can see from equation (11.31) that this perturbation has only one effect and that is to displace the equilibrium point of the oscillator by the distance γ/ω^2. This is physically reasonable since this displacement is exactly what would be expected if a small constant force $m\gamma$ were applied to the mass

m when it is at rest. In that case, the displacement from equilibrium of the mass is

$$\delta x = \frac{m\gamma}{k} = \frac{\gamma}{\omega^2} \qquad (11.34)$$

when the equation for the frequency of the oscillator (11.5) is used.

A second common perturbation is one that varies sinusoidally with the same frequency as that of the oscillator itself:

$$f(t) = m\alpha(t) = m\gamma \cos \omega t$$

so that

$$g(t) = \gamma \cos \omega t \qquad (11.35)$$

where γ is a small constant acceleration. Using the methods developed in the previous case, we can write the differential equations for the quantities $\alpha(t)$ and $\beta(t)$ from equations (11.23) and (11.24) as follows:

$$\frac{d\alpha(t)}{dt} = \frac{\gamma}{\omega} \cos^2 \omega t \qquad (11.36)$$

and

$$\frac{d\beta(t)}{dt} = -\frac{\gamma}{\omega} \sin \omega t \cos \omega t \qquad (11.37)$$

Equations (11.35) and (11.36) can be integrated to yield

$$\alpha(t) = \frac{\gamma}{\omega^2} \int \cos^2 \omega t \, d(\omega t) = \frac{\gamma}{2\omega^2} [\omega t + \sin \omega t \cos \omega t] \quad (11.38)$$

and

$$\beta(t) = -\frac{\gamma}{\omega^2} \int \sin \omega t \cos(\omega t) \, d(\omega t) = \frac{\gamma}{\omega^2} \int \sin \omega t \, d(\sin \omega t)$$

$$= \frac{\gamma}{2\omega^2} \sin^2 \omega t \qquad (11.39)$$

The coefficients $a(t)$ and $b(t)$ that result from this perturbation can now be written as follows:

$$a(t) = a + \frac{\gamma}{2\omega^2}[\omega t + \sin \omega t \cos \omega t] \tag{11.40}$$

and

$$b(t) = b + \frac{\gamma}{2\omega^2}\sin^2 \omega t \tag{11.41}$$

Thus, the perturbed solution is

$$x(t) = a \sin \omega t + b \cos \omega t + \frac{\gamma}{2\omega^2}[\omega t + \sin^2 \omega t + \sin \omega t \cos \omega t] \tag{11.42}$$

Once again, the perturbed solution has the proper form, the sum of the unperturbed solution for the oscillator plus a term caused by the perturbation. In this case, we see that the perturbation has three terms, the first increasing linearly with time and the other two varying periodically with time. Thus, as time increases, the first term begins to dominate, and we therefore have approximately

$$x(t) \approx x_0(t) + \frac{\gamma t}{2\omega} \tag{11.43}$$

where $x_0(t) = a \sin \omega t + b \cos \omega t$ is the unperturbed solution. In this case, what happens is that the displacement of the harmonic oscillator increases linearly with time at a rate determined by the constant $\gamma/2\omega$. This behavior is quite common in both astronomical applications and in the behavior of Earth-orbiting satellites. Linear growth with time such as this is called a "secular" variation.

Finally, let us consider the case in which the perturbing force varies sinusoidally with a frequency different from the natural frequency of the oscillator. The perturbing force is now

$$f(t) = m\gamma \sin \omega' t \tag{11.44}$$

The differential equations for $\alpha(t)$ and $\beta(t)$ now become

$$\frac{d\alpha(t)}{dt} = \frac{\gamma}{\omega} \cos \omega t \sin \omega' t \tag{11.45}$$

and

$$\frac{d\beta(t)}{dt} = -\frac{\gamma}{\omega} \sin \omega t \sin \omega' t \tag{11.46}$$

To integrate these equations, we use the half-angle formulas to evaluate the trigonometric functions:

$$\sin \omega' t \cos \omega t = \tfrac{1}{2}\sin(\omega + \omega')t + \tfrac{1}{2}\sin(\omega - \omega')t$$

$$\sin \omega' t \sin \omega t = \tfrac{1}{2}\cos(\omega - \omega')t + \tfrac{1}{2}\cos(\omega + \omega') \tag{11.47}$$

Now let

$$\omega_1 = \omega + \omega' \quad \text{and} \quad \omega_2 = \omega - \omega' \tag{11.48}$$

Therefore, the differential equations become

$$\frac{d\alpha(t)}{dt} = \frac{\gamma}{2\omega}(\sin \omega_1 t + \sin \omega_2 t) \tag{11.49}$$

and

$$\frac{d\beta(t)}{dt} = \frac{\gamma}{2\omega}(\cos \omega_2 t - \cos \omega_1 t) \tag{11.50}$$

Integrating these equations yields

$$\alpha(t) = -\frac{\gamma}{2\omega}\left(\frac{\cos \omega_2 t}{\omega_2} + \frac{\cos \omega_1 t}{\omega_1}\right) \tag{11.51}$$

and

$$\beta(t) = \frac{\gamma}{2\omega}\left(\frac{\sin \omega_2 t}{\omega_2} - \frac{\sin \omega_1 t}{\omega_1}\right) \tag{11.52}$$

Therefore, the final solution for the perturbed oscillator in this case becomes

$$x(t) = a_0 \sin \omega t + b_0 \cos \omega t$$

$$+ \frac{\gamma}{2\omega} \left(\frac{\sin \omega_2 t}{\omega_2} - \frac{\sin \omega_1 t}{\omega_1} - \frac{\cos \omega_2 t}{\omega_2} - \frac{\cos \omega_1 t}{\omega_1} \right) \qquad (11.53)$$

The frequencies ω_1 and ω_2 are called "beat" frequencies.

An interesting case results when the frequency of the perturbing force, ω', is close to that of the natural frequency of the oscillator. This means that

$$\omega_1 \gg \omega_2 \qquad (11.54)$$

and so the dominant terms in equation (11.52) are those where the denominator is small. Therefore, we can say that, approximately,

$$x(t) \approx x_0(t) + \frac{\gamma}{2\omega\omega_2} (\sin \omega_2 t - \cos \omega_2 t) \qquad (11.55)$$

where the terms with ω_1 as the denominator have been deleted. This perturbation, therefore, leads to a change with a magnitude dominated by the magnitude of ω_2 and a time variation with the lower of the two beat frequencies (again ω_2) as the dominant term.

We shall now apply the ideas we have developed to an important problem in the dynamics of the solar system, the precession of the perihelion of Mercury. The phenomenon we are discussing is illustrated in Figure 11.2. The picture shows the orbit of Mercury with the Sun at the focus. The perihelion is the point of closest approach of the planet to the Sun, and the aphelion is the point where the planet is farthest away from the Sun. (A word about nomenclature: In the case of satellite orbits around Earth, the closest approach of the satellite is called the perigee—from *geos* for earth—and the most distant is the apogee. In the case of planetary orbits around the Sun, the corresponding names are perihelion and aphelion—from *helios* for the Sun.) From Figure 11.2, we can see what precession of the perihelion means: The elliptic orbit of the planet is fixed in the (x', y') coordinate system, where the orbit is defined by the variables r and f, the true anomaly. There is another coordinate system (x, y) that is used to define how the ellipse behaves. The precession of the

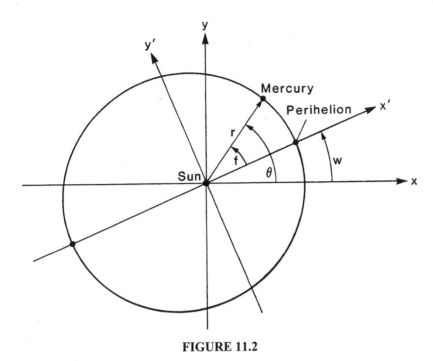

FIGURE 11.2

perihelion of the orbit, or the axis of the ellipse, is defined by the behavior of the angle ω. What is observed is that the angle ω changes slowly as time goes on at a constant rate, $d\omega/dt$. We, therefore, have a secular variation of ω as we have defined it earlier in this chapter.

We will begin by developing an understanding of why the orbit of Mercury slowly precesses around the Sun. Referring to the Table in Appendix II, the eccentricity of the orbit of Mercury is

$$e = 0.2056$$

(Only Pluto has a more eccentric orbit than Mercury.) The orbit of Mercury precesses around the Sun because of the influence of other planets of the solar system, and we will develop a qualitative description of the effect. Referring to Figure 11.3, we will look at the effects of two planets, Venus and Jupiter. Venus has influence because it is closest to Mercury and Jupiter's influence is due to its great mass. On the average, the distance of Venus from Mercury is about 0.65 a.u., the maximum distance between the planets being a little more than one a.u., as shown in Figure

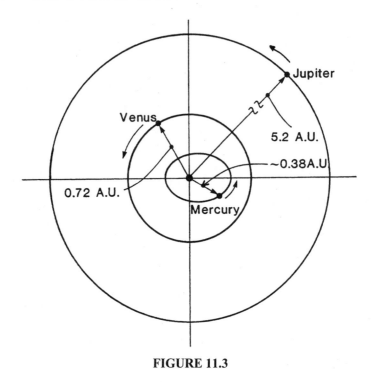

FIGURE 11.3

11.3. The figure of merit of the influence of Venus on Mercury might be estimated as

$$\frac{\text{Mass of Venus}}{(\text{Average distance from Venus to Mercury})^2} \approx \frac{0.97}{(0.65)^2} \approx 2.29$$

Similarly, the same estimate for Jupiter would yield

$$\frac{\text{Mass of Jupiter}}{(\text{Average distance from Jupiter to Mercury})^2} \approx \frac{318}{(5.2)^2} \approx 11.8$$

In making these estimates, we have used the numbers given in Table 10.1. It can be seen that Jupiter has about five times the gravitational influence of Venus on the orbit of Mercury. The effects of the other planets are all substantially smaller than Venus, and so it is a good first approximation to consider only Venus and Jupiter as the cause of the effect.

The first question that must be answered is why the motion of these planets in their orbits should cause the precession that is observed. A qualitative explanation goes something like this: Mercury travels around the Sun once in 88 days, Venus takes 224 days, and Jupiter takes 4333 days to make one circuit around the Sun. Thus, Mercury travels around the Sun 2.5 times for each circuit of Venus and about 50 times for every one circuit by Jupiter. Since Venus moves about twenty times more rapidly with respect to Mercury than Jupiter, it will have a larger effect on the precession of the perihelion of Mercury's orbit even though its gravitational influence is smaller. From the viewpoint of Venus and Jupiter, one could assume that the mass of Mercury is concentrated on a single point on the major axis of Mercury's orbit, as shown in Figure 11.4. This is clearly a crude argument, but it illustrates what happens. It is difficult to calculate the effective center of mass of the orbit, but we can make a rough estimate. Because of Kepler's second law—the law of equal areas—Mercury spends more time on the right side of the orbit than on the left. This happens because Mercury moves more slowly when it is farther away from the Sun. A rough measure of just where the *effective center of mass* of the orbit is can be obtained by looking at the velocity of Mercury at perihelion and aphelion. From the conservation of angular momentum and the notation in Figure 11.4, we have

$$(a - ae)v(\text{perihelion}) = (a + ae)v(\text{aphelion}) \qquad (11.56)$$

so that

$$\frac{v(\text{perihelion})}{v(\text{aphelion})} = \frac{1 + e}{1 - e} = \frac{1.2056}{0.794} = 1.52 \qquad (11.57)$$

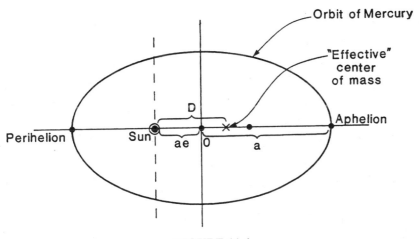

FIGURE 11.4

and therefore

$$v(\text{perihelion}) = 1.52v(\text{aphelion}) \qquad (11.58)$$

Thus, roughly speaking, the time spent on the left side of the Sun compared to the right side of the Sun should be in inverse proportion to the average velocities on the left and the right. Approximately, then, we have

$$\frac{t(\text{left})}{t(\text{right})} \approx \frac{v(\text{aphelion})}{v(\text{perihelion})} \approx 1.52 \qquad (11.59)$$

On the average, then, Mercury spends about 50% more time on the right side than on the left. The effective center of mass of the orbit is therefore some distance D to the right side of the Sun. The calculation of the actual value of D is complicated. The purpose here has been to make a plausibility argument that the effective center of mass of Mercury's elliptic orbit is located at a point different from that of the center of mass of the Sun.

From the viewpoint of Venus and Jupiter, the Sun–Mercury system might therefore look something like the dumbbell shown in Figure 11.5. The *effective center of mass* of the orbit of Mercury can be considered as being "rigidly" attached to the Sun. The motion of Venus and Jupiter affects the "dumbbell" by pulling the effective center of mass of the orbit around the Sun. It is this effect that causes the precession of the perihelion of the orbit of Mercury around the Sun. The precession will be very slow because the Sun's influence is very much larger than that of Venus and Jupiter. If we use the same *figure of merit* that we used to compare the influence of Jupiter and Venus on Mercury, then, for the Sun, we have

$$\frac{\text{Mass of the Sun}}{(\text{Average distance of Mercury from the Sun})^2} \approx \frac{3.32 \times 10^5}{(0.39)^2}$$

$$\approx 2.2 \times 10^6 \qquad (11.60)$$

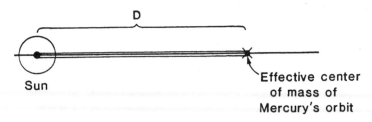

FIGURE 11.5

This number is to be compared with 11.8, which has previously been calculated for Jupiter. The ratio is about 200,000.

In 1865, the astronomer U. Leverrier calculated the secular precession of the perihelion of the orbit of Mercury to be 537 seconds of arc per century caused by the influence of Venus, Jupiter and the other planets in the solar system. To do this required a very complex calculation, but we have outlined the essential principles that were employed to do the job. He found that Venus was responsible for 277 arcseconds of the precession and Jupiter for 153 arcseconds, for a total of 82% of the effect between them. Even though the gravitational influence of Venus is smaller than that of Jupiter, it moves much more rapidly with respect to Mercury. Thus, Venus accounts for about half the calculated precession of Mercury's orbit. Measurements show that the actual precession is 574 seconds of arc per century, and the measurements are accurate enough so that the difference had to be explained. In a classic application of perturbation theory, Albert Einstein showed that the additional 37 seconds of arc could be accounted for by an effect caused by his theory of relativity. It turns out that it is somewhat easier to make this calculation than the one performed by Leverrier. We shall therefore reproduce it here.

The difference in the precession of Mercury's orbit between the measured value and Leverrier's calculated value is caused by the fact that the "mass" of an object is, according to the theory of relativity, a function of the velocity with which the body is moving. The relationship is

$$m = \frac{m_0}{\sqrt{1 - v^2/c^2}} \tag{11.61}$$

where m_0 is called the *rest mass* of the object. Equation (11.61) says that the faster an object moves (that is, the larger the velocity v), the heavier it becomes. It also says that no object can travel more rapidly than the velocity of light. Referring back to Figure 11.4, the mass of Mercury would therefore be a function of where the planet is in the orbit. On the left side of the orbit, where the velocity is larger, the mass of Mercury is slightly larger than the "average" mass, and on the right side, conversely, it is slightly smaller. Since the angular momentum of Mercury in orbit must be conserved, the slight mass changes dictated by equation (11.61) must be compensated for by slight changes in the velocity. These small changes give rise to precession of the perihelion of Mercury's orbit in the same direction as that caused by the influence of Jupiter.

In order to calculate the precession quantitatively, we will apply the perturbation theory developed earlier in this chapter. Referring back to Chapter 4, the equation that leads to the elliptic orbit of Mercury is

$$\frac{d^2u}{d\theta^2} + u = \frac{GMm^2}{L^2} \tag{11.62}$$

where

$$u = \frac{1}{r} \tag{11.63}$$

L is the angular momentum of the planet in its orbit, M is the mass of the Sun, and m is the mass of Mercury. Equation (11.62) is similar to the harmonic oscillator equation discussed earlier so that the same techniques should apply. The solution of this equation is

$$u = \frac{GMm^2}{L^2} + A(\cos f)$$

or

$$u = \frac{GMm^2}{L^2} + A[\cos(\theta - \omega)] \tag{11.64}$$

where the angles θ and ω are defined in Figure 11.2. This equation can be rewritten as

$$u = \frac{GMm^2}{L^2} + [1 + e\cos(\theta - \omega)] \tag{11.65}$$

where e is the eccentricity of the orbit.

We will now develop a perturbation algorithm similar to the one for the harmonic oscillator. First, we define a new variable:

$$v = \frac{du}{d\theta} \tag{11.66}$$

and also

$$\frac{du}{d\theta} + u = \frac{GMm^2}{L^2} \tag{11.67}$$

Equations (11.66) and (11.67) are now modified by a perturbation function $f(\theta)$, and this is done by rewriting equation (11.67) as follows:

$$\frac{du}{d\theta} = -u + \frac{GMm^2}{L^2} + f(\theta) \tag{11.68}$$

Note that we write $f(\theta)$ as a function of θ. It could also be a function of r, because θ and r are related through equation (11.64). We now develop the solutions of equations (11.66) and (11.67), which represent the unperturbed system:

$$u = \frac{GMm^2}{L^2}[1 + e \cos(\theta - \omega)] \tag{11.69}$$

and

$$v = -e\frac{GMm^2}{L^2}\sin(\theta - \omega) \tag{11.70}$$

Now, in order to estimate the effect of the perturbation $f(\theta)$, we assume that both of the "constants" e and ω are functions of the angle θ. Taking the derivative of u with respect to θ yields

$$\frac{du}{d\theta} = \frac{GMm^2}{L^2}\left[\frac{de}{d\theta}\cos(\theta - \omega) - e\left(1 - \frac{d\omega}{d\theta}\right)\sin(\theta - \omega)\right] \tag{11.71}$$

and equation (11.70) becomes

$$v = \frac{du}{d\theta} = -e\frac{GMm^2}{L^2}\sin(\theta - \omega)$$

$$= \frac{GMm^2}{L^2}\left[\frac{de}{d\theta}\cos(\theta - \omega) - e\left(1 - \frac{d\omega}{d\theta}\right)\sin(\theta - \omega)\right] \tag{11.72}$$

Now, equation (11.72) becomes

$$\frac{dv}{d\theta} = -\frac{GMm^2}{L^2}\left[\frac{de}{d\theta}\sin(\theta - \omega) + e\left(1 - \frac{d\omega}{d\theta}\right)\cos(\theta - \omega)\right] \tag{11.73}$$

Thus, substituting (11.73) into (11.68), we have

$$-\frac{GMm^2}{L^2}\left[\frac{de}{d\theta}\sin(\theta-\omega)+e\left(1-\frac{d\omega}{d\theta}\right)\cos(\theta-\omega)\right]$$

$$=-\frac{GMm^2}{L^2}[1-e\cos(\theta-\omega)]+\frac{GMm^2}{L^2}+f(\theta) \qquad (11.74)$$

and this can be rewritten as

$$-\frac{de}{d\theta}\sin(\theta-\omega)+e\frac{d\omega}{d\theta}\cos(\theta-\omega)=\frac{L^2}{GMm^2}f(\theta) \qquad (11.75)$$

Now, equation (11.72) can be rewritten as

$$\frac{de}{d\theta}\cos(\theta-\omega)+e\frac{d\omega}{d\theta}\sin(\theta-\omega)=0 \qquad (11.76)$$

The set of equations (11.75) and (11.76) are a simultaneous pair that can be rewritten as follows:

$$\frac{de}{d\theta}=-\frac{L^2}{GMm^2}f(\theta)\sin(\theta-\omega)$$

$$ \qquad (11.77)$$

$$\frac{d\omega}{d\theta}=-\frac{1}{e}\frac{L^2}{GMm^2}f(\theta)\cos(\theta-\omega)$$

Now, in order to solve equations (11.77), we need to look at the function $f(\theta)$. The formulation of $f(\theta)$ requires us to look at the theory of relativity. Let us take a look at the origin of equation (11.62). Originally, the equation was derived in Chapter 4 from the following equation in terms of r and θ:

$$m\left[\frac{p^2}{m^2r^2}\frac{d^2r}{d\theta^2}-\frac{2L^2}{m^2r^5}\left(\frac{dr}{d\theta}\right)^2\right]-mr\left(\frac{L^2}{m^2r^4}\right)=-\frac{GMm}{r^2} \qquad (11.78)$$

The term on the left is the inertial force, and that on the right is the gravitational force. The small relativistic perturbation arises from the fact that the masses on the left and the right sides of equation (11.78) now are functions of the variables r and θ, because the mass of the planet Mercury changes slightly depending on where it is in orbit. Thus, a general perturbing function $F(r, \theta)$ is added to the left side of equation (11.78), which is derived from the theory of relativity using the same relationships that Albert Einstein used to derive equation (11.61). In terms of the function $f(\theta)$, or $f(r, \theta)$, defined as the "generic" perturbation in equation (11.68), we have

$$F(r, \theta) = \frac{L^2}{mr^2} f(r, \theta) \tag{11.79}$$

Now, $f(r, \theta)$ can be approximated as follows: The theory of relativity calls for an expansion in terms of the variable r, and the first effective term is the one that behaves as r^{-2}. It turns out that the function to be used in equation (11.77) is

$$f(r, \theta) = \frac{3GMm^2}{c^2} \frac{1}{r^2} = \frac{3GMm^2}{c^2} u^2 \tag{11.80}$$

where c is the velocity of light. Note, therefore, that $F(r, \theta)$, defined in equation (11.79) behaves as r^{-4}. Thus, it differs from the gravitational term on the left side of equation (11.78) as it must. This is however only the first term of a series expansion that arises from the theory of relativity.

We can now rewrite the set of equations (11.77) as follows using (11.80):

$$\frac{de}{d\theta} = -\left(\frac{3GMm^2}{c^2}\right)\left(\frac{GMm^2}{L^2}\right)[1 + e\cos(\theta - \omega)]^2 \sin(\theta - \omega)$$

$$\tag{11.81}$$

$$\frac{d\omega}{d\theta} = \frac{1}{e}\left(\frac{3GMm^2}{c^2}\right)\left(\frac{GMm^2}{L^2}\right)[1 + e\cos(\theta - \omega)]^2 \cos(\theta - \omega)$$

We now need to make some assumptions, because equation (11.81) cannot be solved analytically. The fundamental assumption of perturbation theory is that the quantities, e and ω in this case, are slowly varying functions of θ. This is quite reasonable by looking at the coefficients in equation (11.81), since the square of the velocity of light (a very large number) appears in the denominator of the coefficients. Thus, the right sides of equation (11.81) are small by definition, and hence, e and ω are both slowly varying functions of θ.

Evaluating the coefficients yields

$$\left(\frac{3GMm^2}{c^2}\right)\left(\frac{GMm^2}{L^2}\right) = 8 \times 10^{-8} \text{ and}$$

$$\frac{1}{e}\left(\frac{3GMm^2}{c^2}\right)\left(\frac{GMm^2}{L^2}\right) = 4 \times 10^{-7} \tag{11.82}$$

Both of the coefficients are small, as was expected. We can now integrate equation (11.82) for one revolution of Mercury around the orbit. These integrals will yield the *change* in the values of the quantities e and ω for one orbit:

Δe(one orbit) =

$$-\left(\frac{3GMm^2}{c^2}\right)\left(\frac{GMm^2}{L^2}\right)\int_0^{2\pi}[1 + e\cos(\theta - \omega)]\sin(\theta - \omega)\,d\theta \quad (11.83)$$

The integral on the right side of the equation can be rewritten as follows:

$$\frac{1}{e}\int_0^{2\pi}[1 + e\cos(\theta - \omega)]^2 d[e\cos(\theta - \omega)]$$

$$= \left[\frac{1}{3e}(1 + e\cos(\theta - \omega))^3\right]_0^{2\pi} = 0 \quad (11.84)$$

Therefore, the eccentricity of Mercury's orbit does not change under the assumptions that are inherent in the perturbation approximation used here. Integrating the second equation in the set (11.81) yields

$\Delta\omega$(one orbit)

$$= \frac{1}{e}\left(\frac{3GMm^2}{c^2}\right)\left(\frac{GMm^2}{L^2}\right)\int_0^{2\pi}[1 + e\cos(\theta - \omega)]^2\cos(\theta - \omega)\,d\theta \quad (11.85)$$

The integral on the right side of equation (11.85) can be rewritten as follows:

$$\int_0^{2\pi}[1 + 2e\cos(\theta - \omega) + e^2\cos^2(\theta - \omega)]\cos(\theta - \omega)\,d(\theta - \omega)$$

$$= \int_0^{2\pi}[\cos(\theta - \omega)\,d(\theta - \omega) + 2e\cos^2(\theta - \omega)\,d(\theta - \omega)$$

$$+ e^2\cos^3(\theta - \omega)\,d(\theta - \omega)] \quad (11.86)$$

The first and third terms of the integral lead to zero when the integral is evaluated. The middle term yields

$$\int_0^{2\pi}2e\cos^2(\theta - \omega)\,d(\theta - \omega) = \left[e(\theta - \omega)\frac{\sin 2(\theta - \omega)}{2}\right]_0^{2\pi} = 2\pi e \quad (11.87)$$

Therefore, the change in the angle ω during one revolution of Mercury's orbit is

$$\Delta\omega = 2\pi\left(\frac{3GMm^2}{c^2}\right)\left(\frac{GMm^2}{L^2}\right) = 5 \times 10^{-9} \text{ rad/rev} \qquad (11.88)$$

If the result of equation (11.88) is converted to seconds of arc per century, we get

$$\Delta\omega = 43 \text{ seconds of arc/century} \qquad (11.89)$$

This is very close to the difference between the calculated and the measured value of the precession of the perihelion of Mercury.

One of the problems related to the long-time behavior of orbiting bodies is stability, which is an important, difficult, and unsolved problem of celestial mechanics. The stability of the rigid-body (rotational) motion of satellites and space stations is of considerable interest, and it is treated in detail in the literature (see Fitzpatrick, Chapter 14, and Thomson, Chapters 5–7). The following remarks will concentrate on the orbital stability of natural and artificial bodies.

The reader's attention is directed to the over 50 definitions of various kinds of stability, often leading to contradictory conclusions. It is strongly recommended that prior to any announcement of the stability of a dynamical system, the definitions used be clarified. (The problem of the "stability" of the solar system is treated in Chapter 13.)

An elliptic two-body orbit is generally considered stable since a small change of the initial conditions will not markedly change the orbital elements or the shape and orientation of the orbit. This kind of stability, known as orbital stability, is intuitively clear and can be shown analytically without much difficulty. Consider, on the other hand, the same elliptic orbit, and let us change again slightly the initial conditions so that the semimajor axis will change ever so slightly. The orbital stability is still valid, but the change of the length of the semimajor axis will result in a change of the period. After the change of the initial conditions, the small change in the mean motion will displace the body along the orbit. After a sufficiently large number of revolutions, the disturbed body might be close to apogee at the time the body on its original orbit will be at perigee. The two orbits will be very close, but the distance between the body on the original orbit and the body on the slightly changed orbit will be the length of the major axis. This behavior certainly cannot be considered stable in spite of the fact that the original and the new orbits are close. Poincaré (using the first, geometric idea) calls our motion stable, while Lyapunov (using the second, kinematic idea) considers the motion unstable. Poincaré's definition is known as orbital stability, and Lyapunov's is usually referred to as isochronous (equal-time or simultaneous) stability.

An unsolved and rather important problem in celestial mechanics is the stability of the solar system. After the previous example, readers will require a definition of stability before attempting to find the answer. One of the generally accepted definitions of the stability of planetary systems was offered by Laplace (1773), according to which stability requires that the semimajor axes of the planetary orbits show no secular changes, only small periodic changes so that orbits do not intersect. Another similar definition connects planetary stability with no collisions and no escapes.

During the existence of the solar system (estimated at 5×10^9 years) apparently, and probably, no major changes occurred, and numerical integrations indicate stability for the next 10^8 years. Problems with the convergence of analytical series solutions are presently being clarified, but as humiliating as this is, we must admit that the problem of stability is still unsolved (see Roy's book, Chapter 8).

Concerning the details of the second problem of this chapter, see U. J. J. Leverrier, "Théorie du movement de Mercure," *Ann. Observ. Imp. Paris (Mém.),* Vol. 5, pp. 1–196, 1859, and N. T. Roveveare, *Mercury's Perihelion. From Leverrier to Einstein,* Clarendon Press, Oxford (1982). Regarding details of general perturbation methods, see Brouwer and Clemence (1961). A collection of over 50 definitions of stability is given in V. Szebehely, "Review of Concepts of Stability," *Celestial Mechanics,* Vol. 34, pp. 49–64, 1984. The original papers concerning Cowell's and Encke's methods are by P. H. Cowell and A. D. Crommelin, *Mon. Not. Roy. Astron. Soc.,* Vol. 68, p. 576, 1908, and by J. F. Encke, *Berliner Jahrbuch,* 1857. P. A. Hansen's method is described in P. Musen's article in the *Astronomical Journal,* Vol. 63, p. 426, 1958. For a general discussion, see Danby's book, pp. 230–238.

PROBLEMS

11.1. Albert Einstein showed that the perihelion of the orbit of Mercury precesses more rapidly than it would if the theory of relativity were ignored. Without considering relativistic effects, the perihelion of the orbit of Mercury precesses in the same direction as the orbital motion at a calculated rate of 531 seconds of arc per century. This precession is caused by gravitational action of the other planets of the solar system on Mercury. Including relativistic effects, the rate of precession is increased by 43 seconds of arc to 574 seconds of arc, which is the measured value. This means that the relativistic effect also creates an additional precession in the same direction as the orbital motion. The special theory of relativity predicts that

$$m(v) = \frac{m_0}{\sqrt{1 - v^2/c^2}}$$

where $m(v)$ is the mass of the planet Mercury as a function of the linear velocity v of Mercury in its orbit. The quantity m_0 is the rest mass of Mercury, that is, the mass when $v = 0$, and c is the velocity of light, about 300,000 km/s. Using this formula, show that the perihelion of Mercury precesses in the same direction as the orbital motion due to the relativistic mass change.

11.2. Use the formulas in the first problem to show that the difference between the velocity at perihelion and that at aphelion is

$$v_P - v_A = 2\bar{v}\epsilon \left(1 + \frac{\bar{v}^2}{c^2} \right)$$

where

$$\bar{v} = \tfrac{1}{2}(v_P + v_A)$$

The quantity ϵ is the eccentricity of the planet's orbit. From the fact that the relativistic effect adds 43 seconds of arc per century to the precession of the perihelion of the orbit of Mercury, estimate what the relativistic effect adds to the precession of the perihelion of Earth's orbit.

CHAPTER 12

THE MOTION OF
EARTH-ORBITING SATELLITES

In the previous chapter, we examined a case of perturbation theory applied to the motion of a planet (Mercury) orbiting the Sun. In this chapter, we shall look at satellites orbiting Earth and how perturbation approximations can be used to explain their behavior. Before doing this, we will want to define clearly the variables used to describe the motion. Figure 12.1 illustrates the general situation. The motion of an Earth-orbiting satellite is usually referred to the equatorial plane of Earth. The orbital plane defines the plane in which the satellite moves. The angle between the orbital plane and the equatorial plane is called the angle of inclination i. The line of intersection between the orbital plane and the equatorial plane is called the *line of nodes*. In the case of Earth-orbiting satellites, it is often necessary to define an external coordinate system (or at least a direction) with respect to the line of nodes that we will want to use to define how the line of nodes behaves. Normally, we will choose a direction G that points toward the Sun. Having done this, there is an angle Ω between the external direction G and the line of nodes, which is another parameter we will use to define the motion of Earth-orbiting satellites.

Let us now refer to Figure 12.2, which shows the orbit of a satellite moving around Earth in the orbital plane. There are three parameters that define this orbit, the angle ω between the line of nodes and the major axis of the ellipse, a, and the eccentricity of the elliptic orbit, e. The five parameters that have been defined are called the *orbital elements* and these are summarized in Table 12.1. We will see that the orbital elements we

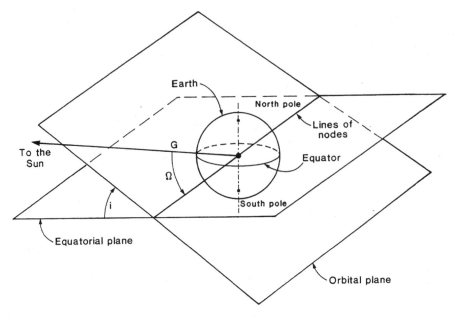

FIGURE 12.1

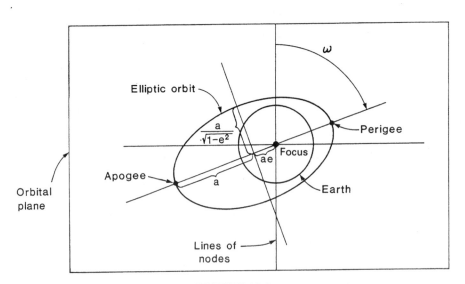

FIGURE 12.2

TABLE 12.1. Orbital Elements

Ω	Angle between G and the line of nodes
i	Angle of inclination between the orbital plane and the equatorial plane
ω	Angle between the major axis of the ellipse and the line of nodes
a	Semimajor axis of the ellipse
e	Eccentricity of the ellipse

have defined completely describe the behavior of a satellite in elliptic orbit around Earth.

In addition to defining the parameters used to describe the orbit, it will also turn out to be useful to review potential theory, because we will have to use methods based on this idea. We will start once again by defining the gravitational force,

$$\mathbf{F}_g = -\frac{GmM}{r^2}\hat{r} \tag{12.1}$$

and the gravitational potential is

$$V_g = \frac{GmM}{r} \tag{12.2}$$

The relationship between the force and the potential is

$$\mathbf{F}_g = \mathbf{grad}\ V_g \tag{12.3}$$

It is of some interest to take a look at the "gradient" operator, which relates the potential function to the force. If the potential function is spherically symmetric, then the gradient operator is simple because it depends only on the variable r:

$$\mathbf{grad} = \hat{r}\frac{\partial}{\partial r} \tag{12.4}$$

where $\hat{r}$ is the unit vector in the r direction. If the potential function also depends on the angles θ and φ, then the gradient operator is more complex. The expression is

$$\mathbf{grad} = \hat{r}\frac{\partial}{\partial r} + \frac{\hat{\theta}}{r}\frac{\partial}{\partial \theta} + \frac{\hat{\varphi}}{r \sin \theta}\frac{\partial}{\partial \varphi} \qquad (12.5)$$

where the angles θ and φ are defined in the polar coordinate system shown in Figure 12.3. The importance of defining these variables has to do with the fact that the actual potential function of Earth is not in fact spherically symmetric.

The shape of Earth is determined by a number of different factors. The most important one is that Earth is not a sphere but rather an oblate spheroid. This means that the diameter of Earth at the equator is somewhat larger than the diameter through the poles. This effect is caused by Earth's rotation and its elasticity, which leads to a centrifugal force that creates a small "bulge" at the equator, as illustrated in Figure 12.4. The exact dimensions are

$$AO = 6356.751 \text{ km}$$

and

$$DO = 6378.136 \text{ km}$$

The difference between the two is

$$DO - AO = 21.385 \text{ km}$$

or about 0.336%. It turns out that even this very small difference has a measurable effect on satellite orbits. Another nonspherical asymmetry is

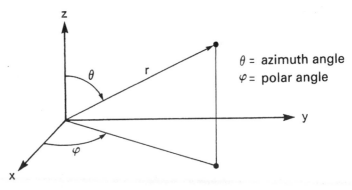

FIGURE 12.3

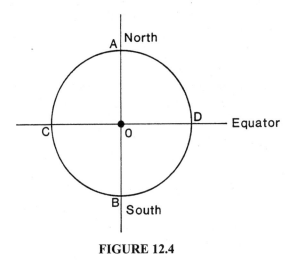

FIGURE 12.4

due to the fact that the equator is not a perfect circle. Referring to Figure 12.5, when we look down on the north pole, we have

$$\frac{NE - NH}{NE} \approx 10^{-5}$$

so that the difference between the largest and the smallest diameter is about 64 m.

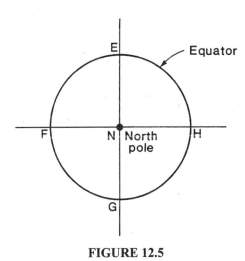

FIGURE 12.5

In addition to these nonsphericities, there are also the so-called mass concentrations. Earth's internal structure consists of a number of layers. These layers are not of uniform density in all three directions, radial, azimuthal, and polar. If the density of the body is constant, then the shape is the essential factor. Deviation from the spherical shape (for uniform density) may be approximated by ellipsoidal bodies, and this is usually the approach for natural celestial bodies. Since the gravitational potential for ellipsoids (of uniform mass distribution) may be represented by Legendre polynomial expansions, this technique has become the most popular approach in describing the gravitational properties of bodies. The approach used in geodesy today is straightforward but not necessarily the final word, and improvements in the basic approach and in the principles used might be forthcoming in the near future. The present method is to write out the potential using Legendre polynomials and, from satellite observations, to evaluate the coefficients in the expansion.

Another approach is known as the inverse problem of celestial mechanics. Usually a force field is given, and for a set of initial conditions, we are establishing an orbit, often by numerical integration. This problem in celestial mechanics might be called the direct problem.

If the orbit is given by observations, we might inquire about the force field that produced this orbit. This is known as the inverse problem, and it is considerably more complicated than the direct problem, for several reasons. It can be shown, for instance, that the force field is not uniquely determined from an orbit, or in other words, there are several force fields that can produce the same orbit. If the general functional form of the force field, or of the potential, is given (such as in the case of the previously mentioned Legendre expansion), then the orbit might be used to determine the coefficients in the expansion. The problem with this "predetermined potential" approach is that the body, such as Earth, usually is unaware of our selection of the functional form of its potential, and when the coefficients in our assumed series approximation are evaluated, they often show dependence on the orbit. This is especially true for the higher order gravitational coefficients of the usual Legendre series approach.

It might be concluded that the inverse problem of celestial mechanics, often referred to as geodesy, is an unsolved problem, and at the present, only approximate analytical descriptions are available for the potential functions of natural bodies.

Regarding the shape of Earth, we use the ellipsoidal approximation and then the corresponding infinite series expansion for the gravitational potential. Consider a spherical Earth first with homogeneous mass distribution, that is, with constant density. An alternate formulation of the basic description is to consider Earth made of concentric spherical shells of uniform density. For these models, as mentioned before, the potential is identical with a point mass potential.

In Chapter 3, we proved the theorem regarding the location of the gravitational center of mass of a spherically symmetric mass. The expression we used for the gravitational potential was

$$V_g(r, \theta, \varphi) = \int_{mass} \frac{Gm\ dM}{r} \tag{12.6}$$

Equation (12.4) can be used to perform the calculations described in the foregoing paragraphs. The potential function V_g can be obtained evaluating the integral. For the present purpose, we will look only at a simple case of an ellipsoid with an equatorial bulge. In that case, the potential function can be expressed as a series of Legendre functions:

$$V_g = \frac{GM_E}{r}\left[1 - \sum_{l=1}^{\infty} \frac{J_l R_E^l}{r^l} P_l(\cos\theta)\right] \tag{12.7}$$

The Legendre polynomials in this expression are defined as follows:

$$P_0 = 1 \qquad P_1 = \cos\theta$$

$$P_2 = \tfrac{3}{2}\cos^2\theta - \tfrac{1}{2} \qquad P_3 = \tfrac{5}{2}\cos^3\theta - \tfrac{3}{2}\cos\theta$$

In the case of a spheroidal distribution, the odd functions drop out because the mass distribution is symmetric in the angle θ. In Equation (12.5), R_E is the average radius of Earth, and r is the distance of the test mass from the center of mass of the spheroid. The coefficients, J_l determine the degree of deviation from a precise spherical shape. It turns out that the only important term is the one that corresponds to $l = 2$, and for this we have

$$J_2 = 1.0826 \times 10^{-3} \tag{12.8}$$

which yields the deviation from sphericity we have already mentioned. The potential function is therefore

$$V_g = \frac{GM_E}{r}\left[1 - J_2\left(\frac{R_E}{r}\right)^2\left(\frac{3}{2}\cos^2\theta - \frac{1}{2}\right)\right] \tag{12.9}$$

We can now employ exactly the same methods that we have developed previously in Chapter 11, where the potential function could also be written as the sum of the unperturbed potential function GM_E/r and a small perturbing term, just as in equation (11.32):

$$V_g = V_0(r) + \Delta V(r, \theta) \tag{12.10}$$

where

$$\Delta V(r, \theta) = -\frac{GM_E J_2}{r}\left(\frac{R_E}{r}\right)^2\left(\frac{3}{2}\cos^2\theta - \frac{1}{2}\right) \tag{12.11}$$

Notice that the term defined in equation (12.11) is small because the coefficient J_2 is small. Also, the perturbing term gets smaller rapidly (like r^{-3}) as the radius of the satellite orbit increases. This is, of course, exactly what would be expected. Thus, the effects on the satellite orbit due to this perturbation are most important for near-Earth-orbiting satellites.

We will not repeat the algebra again and will simply state the results of the calculation:

$$\frac{d\Omega}{dt} = -\frac{3}{2}\frac{2\pi}{T}J_2\left(\frac{R_E}{p}\right)^2\cos i \tag{12.12}$$

and

$$\frac{d\omega}{dt} = \frac{3}{2}\frac{2\pi}{T}J_2\left(\frac{R_E}{p}\right)^2(2 - \tfrac{5}{2}\sin^2 i) \tag{12.13}$$

where T is the period of the satellite in its orbit and

$$p = \frac{L^2}{GM_E m^2} \tag{12.14}$$

which is the semilatus rectum of the orbit of the satellite of mass m. Both equations (12.12) and (12.13) indicate that the perturbation due to the fact that Earth is a prolate spheroid lead to secular variations of the angles Ω and ω. The magnitude of the variation depends upon the angle of inclination between the equatorial plane of Earth and the orbital plane of the satellite. Equations (12.12) and (12.13) can be integrated around one orbit of the satellite to yield the following changes in the angles Ω and ω for each satellite orbit:

$$\Delta\Omega = -3\pi J_2 \left(\frac{R_E}{p}\right)^2 \cos i \qquad (12.15)$$

and

$$\Delta\omega = 3\pi \dot{J_2} \left(\frac{R_E}{p}\right)^2 (2 - \tfrac{5}{2}\sin^2 i) \qquad (12.16)$$

Equation (12.15) says that the line of nodes, which is defined by angle Ω, precesses around the polar axis of Earth. A special case of interest is the sun-synchronous orbit for which the line of nodes is always perpendicular to the line joining the Sun and the Earth ($\Delta\Omega = 2\pi$ per year). These orbits have an inclination of about 94° and are commonly used for Earth observation satellites. Equation (12.16) indicates that there is a critical angle i_c for which the elliptic satellite orbit does not precess:

$$\sin i_c = \frac{2}{\sqrt{5}} \qquad (12.17)$$

and when this is evaluated, it turns out that the angle is

$$i_c = 63°26'5.8''$$

Highly eccentric orbits with this inclination (Molnya orbits) are used for communications satellites by the Russians, because they can provide long and stable residence times at high, northern latitudes.

Another effect that is of interest is one that affects geosynchronous orbits for which the angle of inclination is zero. In this case, the ellipticity of the equator that we have already mentioned has a measurable effect. When perturbation theory is applied to this effect, it is found that a geosynchronous orbit, which is slightly elliptic, has two stable and two unstable points. The stable points are positioned along the minor axis of the ellipse that defines the shape of the equator, and the unstable points are along the major axis (see Figure 12.5). These considerations are important for orbital station keeping, since locating the satellite at the stable points in the orbit minimizes the fuel necessary to maintain the orbit.

In addition to the nonspheroidicity of Earth, there are several other effects that result in deviations from two-body orbits of satellites. It is important to note that these effects depend on the orbit; therefore, during the lifetime of a satellite, the importance of terms in the equations governing its motion could change in time. It should be realized that,

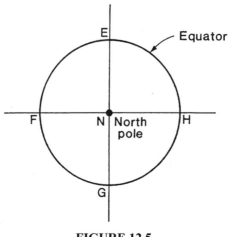

FIGURE 12.5

depending on the accuracy requirements, certain effects can be neglected.

Satellites in equatorial, geosynchronous orbits are primarily affected by the ellipticity of the equator, by their longitude (location), and by solar and lunar perturbations.

If the altitude of the satellite is below 1600 km, the solar and lunar effects (so-called third-body effects) might be neglected.

The atmospheric drag becomes important below 600 km, and it strongly depends on the shape and mass of the satellites. It can be shown that atmospheric drag has a secular effect on the semimajor axis, which decreases, and on the eccentricity, which approaches zero. The orbit becomes more and more circular with decreasing semimajor axis until entry occurs. The precise evaluation of drag effects presents one of the most difficult problems in orbit mechanics with location and time. The unpredictable solar activities and their effects on the density make the entry computations rather uncertain.

The atmospheric drag per unit mass of the satellite is usually evaluated from the equation

$$D = \tfrac{1}{2} c_D \rho V^2 A$$

where c_D is the drag coefficient, ρ is the atmospheric density, V is the velocity, and A is the area of the satellite normal to the velocity vector. The density depends on the location and on the time as mentioned before, but the major effect can be represented by the equation

$$\rho = \rho_0 e^{-\alpha h}$$

where ρ_0 and α are constants and h is the altitude. The area A might also vary since it depends on the attitude of the satellite and, therefore, it can change if the satellite is not spherical. The above formula for the drag might be modified by introducing the ballistic coefficient,

$$B = c_D \frac{A}{m}$$

where m is the mass of the space probe. The density variation is often represented in tabular form (Jacchia) instead of by the above approximate equation. For all the above reasons, numerical integration becomes mandatory to determine the atmospheric effects on low-elevation satellites.

Solar radiation effects are important for low-density (balloon) satellites with large surface area and with small mass.

The previously mentioned deep-space orbits present major problems in celestial mechanics. These high-eccentricity satellites are influenced by atmospheric drag and zonal gravitational harmonics during the time they are near perigee. As these satellites approach their apogee, the perturbations due to atmospheric drag and gravitational perturbations become less important than the lunisolar effects.

The effects of Earth's magnetic field and of collisions with charged and noncharged particles are generally small on the orbital elements.

Collisions with orbital debris, such as parts of inactive satellites, rocket bodies, parts of missiles, and so on, present very serious danger to the functioning of satellites, space stations, and space vehicles in general. At high altitudes, the dispersion of the debris is such that the probability of impact is small. The geosynchronous altitude is an exception since only recently it became mandatory that inactive communicative satellites be removed. At low altitudes, the probability of impact is reduced since the "space garbage" sooner or later reenters the atmosphere and burns up.

Once again, the inverse problem should be mentioned since satellite orbits offer information, besides higher order gravitational coefficients, about the density of Earth's atmosphere. Orbital decay allows the determination of the density distribution and its dependence on the altitude.

In conclusion, a recently proposed technique (1974) using tethered satellites is mentioned. This system consists of a space station, or a space vehicle, such as a space shuttle, which, while in orbit, deploys a small satellite into a region where the atmospheric density is to be measured.

This satellite will not reenter since it is tethered, and from its dynamical behavior, the atmospheric density can be determined. Other possible uses of such tethered satellite systems are to generate electricity using Earth's magnetic field and to allow altitude changes of the shuttle.

The basic references are by Brouwer and Clemence (1961) and by Kovalevsky (1963). Concerning the still intriguing problems of drag and radiation effects, see Battin's (1987), Escobal's (1965), Fitzpatrick's (1970), and King-Hele's (1964) books. A recent report on tethered satellites is by V. R. Bond, NASA-JSC-22681 (1987).

EXAMPLES

12.1. Consider an artificial Earth satellite in a plane at 30° from the Equator ($i = 30°$) with perigee and apogee heights $h_p = 161$ km and $h_a = 837$ km. Neglecting the effects of drag, find the secular variations of the orbital elements.

The semimajor axis and eccentricity are related to h_a and h_p by the well-known equations

$$a(1 - e) = R_e + h_p = r_p$$

and

$$a(1 + e) = R_e + h_a = r_a$$

where R_e is the equatorial radius of Earth. Note that the elevations are denoted by h_p and h_a. The corresponding distances from the center of Earth are r_p and r_a.

The above equations give

$$a = \tfrac{1}{2}(r_p + r_a) \quad \text{and} \quad e = \frac{r_a - r_p}{r_a + r_p}$$

or

$$e = \frac{r_a - r_p}{2a}$$

Note that the second equation for e is simpler than the first, but it contains a previously computed value (a). Formulas using the original inputs (r_a and r_p) have the advantages of showing the functional dependence of the element (e) on the inputs as well as avoiding propagating errors made in the previously computed element (a) as mentioned before. Also note that the above equations may be written as

$$a = R_e + \tfrac{1}{2}(h_a + h_p)$$

and

$$e = \frac{h_a - h_p}{2R_e + h_a + h_p}$$

The semilatus rectum is given by $p = a(1 - e^2)$ or by

$$p = \frac{2r_a r_p}{r_a + r_p}$$

and the ratio needed to evaluate $d\Omega/dt$ and $d\omega/dt$ is

$$\frac{R_e}{p} = \frac{R_e(r_a + r_p)}{2r_a r_p}$$

The mean motion is computed from Kepler's law:

$$\frac{2\pi}{T} = \left(\frac{GM_E}{a^3}\right)^{1/2}$$

or

$$\frac{2\pi}{T} = \left[\frac{8GM_E}{(r_p + r_a)^3}\right]^{1/2}$$

The time derivatives of Ω and ω become

$$\frac{d\Omega}{dt} = -A \cos i$$

and

$$\frac{d\omega}{dt} = A(2 - \tfrac{5}{2} \sin^2 i)$$

where

$$A = \frac{3J_2 R_e^2}{4r_a^2 r_p^2}\sqrt{2GM_E(r_a + r_p)}$$

Note that these equations allow direct use of the original input values. For the numerical values mentioned above, we have

$$a = 6877.034 \text{ km}$$

$$e = 0.04914$$

$$p = 6860.425 \text{ km}$$

$$\frac{R_e}{p} = 0.92970$$

$$2\pi\omega = 1.10705 \times 10^{-3} \text{ rad/s}$$

$$P = 5675.61 \text{ s}$$

$$\frac{d\Omega}{dt} = -1.3457 \times +10^{-6} \text{ rad/s} = -6.66°/\text{day}$$

$$\frac{d\omega}{dt} = 2.1366 \times 10^{-6} \text{ rad/s} = 10.58°/\text{day}$$

The changes of Ω and ω in one revolution of the satellite are $\Delta\Omega = -7.64 \times 10^{-3}$ rad and $\Delta\omega = 12.13 \times 10^{-3}$ rad.

12.2. The artificial satellite *Explorer 6,* known as 1959 δ2, was discussed in Chapter 8, Example 12.7. Using the results obtained there, compute the major secular effects on the node and on the argument of perigee when the inclination of the orbit to the equatorial plane is 45°.

The regression of the ascending node becomes $d\Omega/dt = -4.155 \times 10^{-3}$ rad/day, or 0.238 deg/day, and the progression of the apsidal line is $d\omega/dt = 4.407 \times 10^{-3}$ rad/day, or 0.252 deg/day.

12.3. The semimajor axis and the eccentricity of an artificial Earth satellite are 8676 km and 0.19, respectively, and it is in an orbit inclined to the equatorial plane by 34°. The change of the argument of

perigee is 4.41°/day and that of the nodal line is –3.019°/day. Compare the two values of J_2 that the above observations give.

Solving equations (12.15) and (12.16) for J_2, we have

$$J_2 = -\frac{2}{3}\frac{T}{2\pi}\frac{d\Omega}{dt}\left(\frac{p}{R_e}\right)^2\frac{1}{\cos i} = 1.07916 \times 10^{-3}$$

and

$$J_2 = -\frac{4}{3}\frac{T}{2\pi}\frac{d\omega}{dt}\left(\frac{p}{R_e}\right)^2\frac{1}{4 - 5\sin^2 i} = 1.07275 \times 10^{-3}$$

Note that the difference between these two computed values is in
the sixth decimal and that the presently accepted value of $J_2 =$
1.08263×10^{-3} shows a five-figure agreement with our value computed from $d\Omega/dt$.

12.4. Find the angles of inclination for which the motions of the nodal
line and of the argument of the perigee are in resonance.

By resonance, we mean that the ratio of the angular velocities of
these motions are rational numbers, or

$$-\frac{d\Omega/dt}{d\omega/dt} = \frac{n_1}{n_2}$$

where n_1 and n_2 are integers. From the above relation, we have

$$n_2\frac{d\Omega}{dt} + n_1\frac{d\omega}{dt} = 0$$

which is the usual way to express resonance conditions. If the right
side of the above equation is only approximately zero, we have near
or almost resonant motion. Substituting in the above equation the
expressions given by equations (12.15) and (12.16), we have

$$k = \frac{n_1}{n_2} = \frac{5\cos^2 i - 1}{2\cos i}$$

where k is any rational number.

All angles of inclinations that satisfy the above equation result
in resonance.

For instance, for $k = 1$, we have $i = 46.378°$, or $i = 106,852°$, and for $k = \frac{1}{2}$, the resonance condition gives $i = 56.065°$, or $i = 110.993°$. (Note that resonance also exists for negative values of k. Regarding $k = 2$, which value gives $i = 0$, see Figure 12.2, where ω was introduced as the proper variable for this special case.)

PROBLEMS

12.1. Jupiter's fifth satellite, Amalthea, shows a precession of the line of apsides, $d\omega/dt = 2.51$ deg/day. The orbit is approximately circular with $a = 181,200$ km, and its inclination may be neglected. Find the oblateness parameter (J_2) of Jupiter.

12.2. Find the oblateness parameter (J_2) of a planet having the same radius and mass as Earth by observing the orbit of an artificial satellite around the planet having pericenter altitude $h_p = 200$ km and apocenter altitude $h_a = 400$ km; inclination from the equatorial plane is $i = 30°$; and pericenter shift is $\Delta\omega = 10$ deg/day.

CHAPTER 13

THE PROBLEM OF THREE BODIES AND THE STABILITY OF THE SOLAR SYSTEM

Everything we have described so far and all of the quantitative relationships that have been developed assume that the problem to be described is dominated by two bodies moving in the gravitational field produced by these bodies. In Chapter 11, we introduced the notion of perturbations by small effects due to "third" bodies that affect the motion of the two-body system. The precession of perihelion of Mercury is caused mostly by a third body, Jupiter. What we want to do now is to take a more general look at the case of more than two bodies in a gravitationally interacting system.

The basic and original definition of the problem of three bodies is as follows: Three point masses (or bodies of spherical symmetry) gravitationally attract each other; for a given set of initial conditions, find the resulting motion.

In addition to the statement of the problem, we can say that the motion takes place in three dimensions, and there are no restrictions concerning the masses, the initial positions, and the initial velocities. Without actually writing down the equations of motion, we can expect three second-order differential equations for the three position vectors of the three bodies, forming a $2 \times 3 \times 3 = 18$th-order system. The energy is conserved since it is a conservative system; the angular momentum is conserved since there are no moments acting, and the center of mass of the three bodies moves with constant velocity. It is not hard to write down the equations of motion for the general problem of three bodies. Once the equations are available, the integrals of energy, angular momentum, and center-of-mass motion can be derived. The methods used to do this are the same as those introduced in Chapter 2. These "integrals" are sufficient to solve the two-body problem, but they do not suffice for the three-body problem, because there are too many variables that have to be considered to solve the problem.

Isaac Newton was one of the first people to investigate the problem of three bodies formed by the Sun, Earth, and the Moon. Later, he recalled that "my head never ached but with my studies of the Moon," because the mathematics involved was so very difficult. He also complained that thinking about the problem kept him awake at night. In spite of these afflictions, Newton was finally able to develop methods for computing the motion of the perigee of the Moon's orbit to within 8% of the observed value (1687).

Following Newton, a number of prominent mathematicians attacked the problem by making approximations that permitted "closed" mathematical solutions. In 1760, Leonhard Euler developed solutions for a special case of the three-body problem in which two of the three bodies are fixed in space and the third moves in their gravitational field. This problem can be solved by elliptic functions, and it is useful because, among other things, this approximation can well represent the motion of an Apollo spacecraft in the gravitational field of Earth and the Moon (see Vinti, 1961). In 1772, Euler also proposed the use of a rotating synodic, coordinate system to approximate, for example, the almost circular motion of the Moon around Earth. This allows more accurate approximations of the motion of bodies in the gravitational field of Earth and the Moon. These ideas have been generalized in what is called the *restricted problem of three bodies.* In this general case, two of the three bodies have a mass that is much larger than the third. As a result, the motion of the two larger masses will not be influenced by the third body, but the larger bodies will govern the motion of the small body. A good example of immediate practical importance is the system consisting of Earth, the Moon, and a space probe traveling on a lunar trajectory. The masses of the participating bodies are approximately in the ratio

$$m_E : m_M : m_P \approx 100 : 1 : 10^{-19}$$

where m_E, m_M, and m_P are the masses of Earth, the Moon, and a 6000-kg probe. If the forces are computed, we might see that the effect of the probe on Earth is always 16 orders of magnitude smaller than the Moon's effect on Earth. Furthermore, the effect of Earth on the Moon is always 16 orders of magnitude larger than the effect of the probe on the Moon. We can continue these relative-force evaluations (using only two-body effects) and find Earth's and the Moon's effect on the probe as it travels from Earth to the Moon. When it is in the vicinity of Earth, the Moon's effect may be neglected since it is 10^{-5} times that of Earth's. When the probe is close to the Moon, Earth's effect is 10^{-3} times that of the Moon's effect.

If the model of the restricted problem is applied to a very large "probe," such as to the Moon as influenced by Earth and by the Sun, the above 10^{-16} neglected order of magnitude becomes 10^{-2}. Therefore, the restricted problem should be used only as a first approximation to establish the orbit of the Moon (see the examples at the end of this chapter).

The essential idea of the restricted problem is that we can separate the motion of the two large bodies (often called the primaries) and solve this two-body problem first without considering the third small body. After the problem of two bodies is solved, we investigate the motion of the small body in the (known) gravitational field of the two large bodies. In many problems of interest in space dynamics, the primaries move on approximate circles (Earth and the Moon, the Sun and Jupiter, etc.). This results in the simplest form of the problem, known as the circular restricted problem of three bodies.

It is important to recognize that the "restricted" problem of three bodies is closely related to perturbation theory. In Chapter 11, we showed that it is possible to calculate small changes in the behavior of a two-body system by using the two-body motion as a first approximation and calculating the small changes in the motion due to a third body, or another small perturbation. In the restricted three-body problem, we consider the two large bodies as fixed with respect to each other, that is, the first-order approximation, and then calculate the motion of the third "small" body in the gravitational field of the other two bodies.

In the late 1770s, the distinguished French mathematician Joseph Louis Lagrange applied his formidable powers to treatments of the restricted problem of three bodies. When the primaries move on circles, he discovered that there are five points in the plane of their motion where the forces acting on a probe are balanced. These forces are the gravitational attractions of the large masses on the probe and the centrifugal force acting on the probe revolving with the primaries. This revolving system is known as the synodic system, in which the primaries are fixed. In a fixed inertial system called the sidereal system, the primaries are moving in circles. Three of the equilibrium points are located on the line connecting the primaries and two points form equilateral triangles with the primaries. The three colinear points are unstable, and the two triangular points are stable for small mass ratios. This means that if outside forces are acting on the probe, which is placed at any of the colinear equilibrium points, the probe will depart. A probe placed at the triangular points will librate instead of escaping, provided the mass ratio of the primaries is smaller than 0.0385. The triangular points, therefore, are also called points of libration. (Libration is an oscillatory motion around an equilibrium point.) Because of their discoverer, the equilibrium points are also known as Lagrangian points. The condition of the stability of the triangular points (i.e., the small mass ratio of the primaries) is satisfied for systems of interest in space dynamics. For instance, for the Earth–Moon system, the value of the mass parameter is 0.012. The instability of the colinear locations might be counteracted by station-keeping propulsion systems. Figure 13.1 shows the locations of the five equilibrium points for the Earth–Moon system.

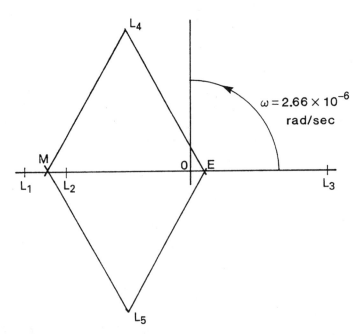

FIGURE 13.1 Location of the equilibrium points in the restricted problem of three bodies.

Some of the numerical values describing this system are

$$m_E = 5.97 \times 10^{24} \text{ kg} \qquad m_M = 7.35 \times 10^{22} \text{ kg}$$

$$EM = 384.4 \times 10^3 \text{ km} \qquad OE \approx 0.012 \times EM$$

$$ML_1 \approx ML_2 \approx 0.16 \times EM \qquad EL_3 \approx EM$$

The first and second colinear libration points (L_1 and L_2) are located at 16% of the Earth–Moon distance, and the third (L_3) is approximately symmetrical to the Moon's location. For the triangular points, we have

$$EL_4 = ML_4 = EL_5 = ML_5 = EM$$

To make clear how these equilibrium points are created, Figure 13.2 shows the gravitational equipotentials that illustrate how the so-called Lagrangian points come about. This diagram shows one way of displaying the combined gravitational fields of two massive objects. Each line represents an equipotential contour, along which a small object would feel the same total force. The resulting contour map serves as a topographic map, showing "hills" and "valleys" in the force field. Any small object in this field will feel a force pulling it in the "downhill" direction.

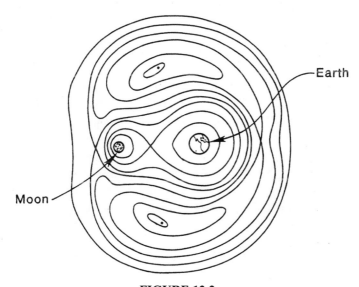

FIGURE 13.2

The basic and important difference between the problems of two and three bodies is that the former is integrable but the latter is not. The somewhat complicated concept of integrability is associated with the names of H. Bruns (1887) and Henri Poincaré (1890), and it usually refers to the availability of generally valid analytical solutions. For any given set of initial conditions, we can predict the quantitative and qualitative behavior of the problem of two bodies. This is not the case for the problem of three bodies. If an analytical solution is desired for a certain set of initial conditions for the problem of three bodies, this solution is usually expressed by infinite series. If the series converges for any length of time, we have an analytic solution for a given set of initial conditions. Changing the initial conditions, the solution might change completely, and the series might diverge. This solution, consequently, has no general validity. Furthermore, solutions given by infinite series (even when they converge) do not offer a qualitative picture. For the problem of two bodies, the initial conditions will tell us the qualitative nature of the solution, such as the type of orbit the bodies will follow: circular, elliptic, parabolic, or hyperbolic. This is not the case in the problem of three bodies, where the qualitative properties of the solution are generally not known.

Another representation of the concept of integrability is that the system has a sufficient number of generally valid and independent invariant relations (integrals) between the variables so that analytical solutions can be obtained. The idea of integrability is not a simple one, and it does not belong to an introductory textbook; nevertheless, its importance warranted the above short description.

As an additional note, the reader is reminded of two, already mentioned, concepts of some interest. First, it is recalled that the time dependence of the variables in the solution of the problem of two bodies cannot be expressed by closed-form functions, as made clear by Kepler's equation. This fact does not mean that this problem is not integrable. The second remark is related to the solvability of a dynamical problem versus its integrability. By solution, we can mean a special solution obtained by numerical integration, valid for a certain given time. Such solutions, of course, exist and might be obtained for the problem of three bodies. As the time increases, the special solution will lose its accuracy and validity. Not so for the problem of two bodies. If the initial conditions indicate an elliptic solution, the participating two bodies will remain on their elliptic orbits forever.

Henri Poincaré's proof that the general problem of three bodies interacting through gravitational forces has no "closed" mathematical solution was only the first step. What was really important is that he then asked himself what statements can be made about the motion of the system even though no analytic solutions exist. By employing the concept of phase space (about which more later), Poincaré developed an ingenious method for tracing the behavior of the three-body problem for many different mass ratios and initial conditions. In performing these calculations, he made the startling discovery that we have already mentioned briefly. For a certain set of initial conditions, the three bodies (as represented by their "motion" in phase space) execute a periodic motion, which can be predicted for all time by the governing differential equations. For a slightly different set of initial conditions, however, the resulting trajectory could be very different. Sometimes, the new set of initial conditions yielded a trajectory that was predictable by the governing equations. At other times, the trajectory Poincaré calculated represented a "chaotic" behavior that could not be predicted by the original equations. He was therefore confronted with the paradox that a system such as the three bodies interacting through a gravitational field could be represented by a completely deterministic set of equations (Newton's gravitational equations in this case) but that in certain instances the behavior of the system could not be predicted.

A simple thought experiment can be used to illustrate this state of affairs. Imagine a rigid rod attached to a bar with a frictionless bearing. When at rest, the rod hangs vertically, but when the bottom end is given a push, the rod starts to move. If the initial push imparts an angular velocity to the rod below a certain critical value, then it will swing as a pendulum. Above that critical value, it will "go over the top" and revolve around the bearing. The difference between the initial conditions, that is, the initial push, which leads to two entirely different outcomes, may be very small. If the rod is replaced by a string with a weight on its end, then there is a third possibility. In addition to pendulum motion and rotation, there are values of the initial push for which the velocity is such that the centrifugal force is too small to maintain the tension in the string when the weight is near the top of the circle. Therefore, the weight falls and the string stops it short when it is stretched again. The subsequent motion is not predictable but rather chaotic. The properties of the three-body problem that we have described are characteristic of all systems governed by nonlinear equations.

In Chapter 10, we used the simple harmonic oscillator to illustrate how perturbation theory works. We shall use the simple harmonic oscillator again to describe in the most elementary terms what happens in the case of nonlinear dynamics. The fundamental result of the application of high-speed computers to the solution of nonlinear differential equations is what

Poincaré had already glimpsed in his early work on the three-body problem. It might be stated as follows: Nonlinear differential equations of the type we are considering here are deterministic. This means that they express a definite relationship between the variables. There are two classes of solutions of such equations. One, called the special solutions, has closed-form expressions, such as the solution of the equations for elliptic and hyperbolic orbits in the case of the two-body problem. However, there is another class of solutions that does not permit a precise prediction of the behavior of the system. In short, these solutions satisfy the equations, but they are not *determined* by them, and as we have seen, these solutions depend sensitively on the initial conditions that are imposed on the system.

The linear harmonic oscillator is the simplest dynamical system. It is governed by the equation

$$m\frac{d^2x}{dt^2} + kx = 0 \tag{13.1}$$

and leads to sinusoidal oscillations, as shown in Figure 13.3. A mass on a spring in a gravitational field behaves as a linear harmonic oscillator. If instead of the distance versus time, or *time series* description shown in the diagram, we adopt a *phase space* with distance along one axis and velocity along the other, then the linear harmonic oscillator describes a circle as shown in Figure 13.4. This circle is called the phase space diagram of the motion. If a dissipative term is added to equation (13.1), then it looks like this:

$$m\frac{d^2x}{dt^2} + l\frac{dx}{dt} + kx = 0 \tag{13.2}$$

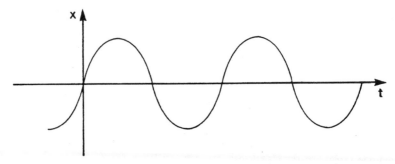

FIGURE 13.3 Sinusoidal oscillation.

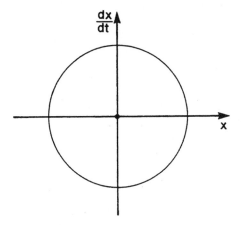

FIGURE 13.4 Phase diagram of a sinusoidal oscillation.

The result of this motion is shown in Figure 13.5 and 13.6. The phase space diagram of the damped oscillator is sometimes called an *attractor*, because the motion causes the point in the phase space as it moves to be attracted to the origin of the coordinate system.

This same idea can be applied to nonlinear motions as well. Take, for example, the nonlinear equation

$$\frac{d^2x}{dt^2} + k\frac{dx}{dt} + x^3 = B\cos t \tag{13.3}$$

We can also plot a time series for this equation that is more complicated than the time series for linear harmonic motion. An example is shown in

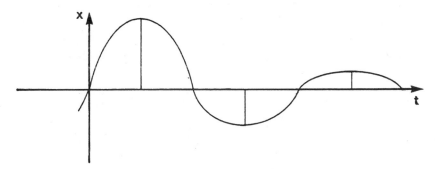

FIGURE 13.5 Oscillation of a damped system.

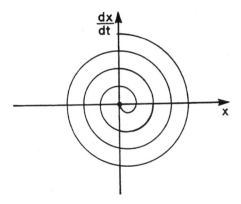

FIGURE 13.6 Phase diagram of a damped oscillator.

Figure 13.7. We can also plot a phase space diagram of the motion, and both of these are shown in Figure 13.8 and 13.9. Instead of circular motion, we now see the more complex diagram, but what is interesting is that the motion in phase space has a very definite "order." Instead of moving randomly in the phase space, the point that describes the dynamical behavior of the system executes specific orbits. These orbits never close on themselves, since the motion is not periodic, but they do have a well-defined pattern. The pattern, furthermore, depends on how the motion was started. In other words, it depends on the *initial conditions* imposed on the equation.

Edward Lorenz, who was one of the first to study nonlinear equations on powerful desktop computers, called these patterns *strange attractors,* and it is the strange attractors that seem to indicate that there is, in fact, an underlying order in the chaotic motions of nonlinear systems. What Lorenz found in exploring the *solution space* of nonlinear equations was that there is a level of order that can be shown to exist only after the capability to perform vast numbers of numerical calculations was in hand. Lorenz showed that the initial conditions of the problem defined a solution space for the equations that exhibited quite regular patterns. Another way of saying this is that the phase space occupied by the solutions of the system was quite limited and exhibited a regular pattern depending on how the calculation was started. Lorenz called these regularities in the solution space of the equation the strange attractors around which the solutions of the nonlinear dynamical system he was considering seemed to congregate.

The appearance of these new "regularities," by the way, gives the lie to

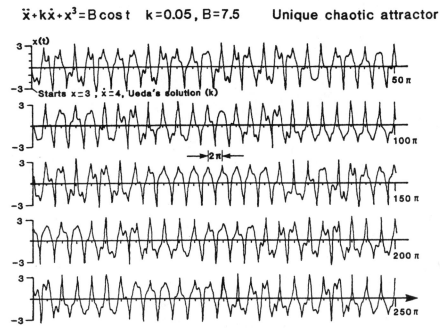

$$\ddot{x} + k\dot{x} + x^3 = B\cos t \quad k = 0.05, \ B = 7.5 \qquad \textbf{Unique chaotic attractor}$$

Starts $x = 3$, $\dot{x} = 4$, Ueda's solution (k)

FIGURE 13.7 Time series of a steady-state response.

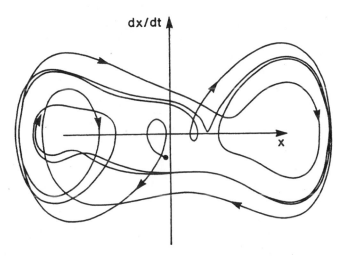

FIGURE 13.8 Steady-state chaotic trajectory in the phase plane $(x, dx/dt)$ of the nonlinear differential equation $d^2x/dt^2 + 0.4\,(dx/dt) - x + x^3 = 0.4\sin t$. This behavior is typical of such nonlinear equations that have complex attractors of this kind.

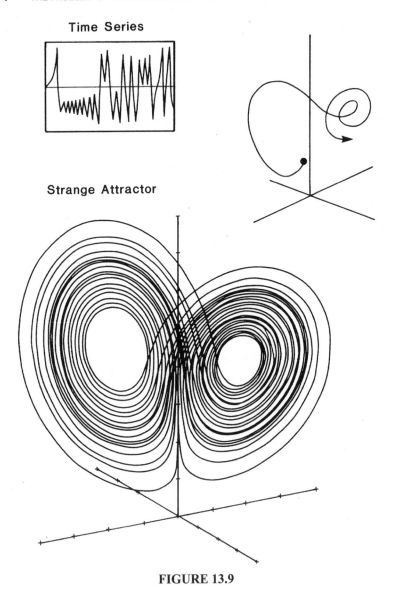

FIGURE 13.9

the statement that computers are only tools, or computational aids. The fact is that brand-new and genuine scientific insights can arise from large-scale computations precisely because they permit us to explore the solution space of a nonlinear system more completely than we have ever been able to do before. If we look at Johannes Kepler's work 300 years ago, we should not be surprised. When Kepler painstakingly used Tycho

Brahe's observations of the orbit of Mars to derive the laws of planetary motion, he really did demonstrate the importance of computational physics! It is, indeed, true that the most important lessons have to be re-learned every once in a while.

We will now return to the problem of three bodies and show how it is related to the stability of the solar system. Following Poincaré's work on the three-body problem, many people worked to establish the region of the parameter space for a given problem in which well-defined closed solutions existed and those for which there were none. In the course of this work, it was discovered that there were chaotic solutions for the three-body problem in which one of the bodies acquired sufficient velocity to escape from the other two. This happens even when the total energy of the three-body system is negative. Poincaré himself was, of course, aware that this could happen, and it was this phenomenon that focused his attention on the problem of the "stability" of the solar system.

If a negative total-energy three-body system "ejects" one of the bodies, it can be said to be unstable. What the word *unstable* means in this case is that the end state of the system (two bodies orbiting around each other and a third one infinitely far away) is qualitatively different from the initial state (three bodies moving in their mutual gravitational fields). Poincaré, therefore, decided to revisit the problem of proving that the solar system is stable.

In the early nineteenth century, Pierre Simon de Laplace used perturbation theory to show that the solar system is stable. What he did was to treat the solar system as a many-body problem in which not only the action of the Sun on the planets but also the effects that the planets have on each other were considered. (We have already illustrated some of these effects in Chapter 11.) He wanted to show that any changes in the motion of the planets due to these perturbations did not lead to secular divergences (i.e., "ejections" of planets in the solar system). Since most of the perturbations caused by one planet on another are periodic in nature, Laplace developed solutions using Fourier series expansions. He succeeded in showing that for the perturbations he considered, the orbits of the planets, to a good approximation, would not be significantly altered. Poincaré studied Laplace's work and discovered that some of the series expansions Laplace had used were not absolutely convergent. Therefore, Laplace's conclusion that the planetary orbits do not change substantially and, therefore, that the solar system is stable was not warranted.

The question of what "stable" means in a system as complex as the solar system needs to be examined more carefully. One definition, as in the case of the three-body problem, might be the requirement that the solar system does not eject a major planet. Recently, very powerful numerical integrating machines have been built that are programmed to integrate the detailed differential equations describing the solar system forward in time. (These are called digital orreries.) Using these machines, it has been demonstrated that the nine major planets execute minor excursions around their current orbits, but there is no indication that the orbit of any planet gains enough energy to be ejected. Nor is there any indication of a major change in the orbital parameters of any of the planets. Since the estimated lifetime of the Sun is about 10 billion years and the age of the solar system is about 4.5 billion years, more integration time on the digital orrery is required to show that the solar system is stable for the entire lifetime of the Sun. Some work has been done already that indicates that if the solar system's major planets exhibit instabilities, then the growth rates are very long, 20–40 billion years. This work has been done by Jack Wisdom and his associates at the California Institute of Technology. See Ivars Peterson's book (1993) in the Appendix.

There are, of course, other ways of defining stability. One could, for example, look for evidence of chaotic motion in the solar system, and if an example is found, then one could perhaps broaden the definition of stability to include phenomena that are characteristic of chaotic motion. An intriguing example is the case of the *Earth-crossing* asteroids. The vast majority of the approximately 5000 known asteroids execute almost circular orbits at an average distance of about 2.5 a.u. from the Sun. A small number, however, are in orbits that are sufficiently eccentric so that they actually spend some time inside the orbit of Earth, hence the term Earth crossing. In 1857, Kirkwood proposed a possible explanation for this observation. He carefully examined the orbits of about 50 large asteroids, and he discovered that asteroids having periods that have a simple fractional relationship with the orbit of Jupiter (e.g., 3 to 1, 5 to 2, 7 to 3, and 2 to 1) are missing. These are the so-called Kirkwood gaps. Asteroids with periods having simple fractional relationships with the period of Jupiter will experience a perturbation of their orbits always in the same position in the orbit. Kirkwood conjectured that the cumulative effect of these perturbations was large enough to throw these asteroids into highly eccentric orbits around the Sun. In 1981, using the methods of chaos theory, Jack Wisdom showed that Kirkwood's conjecture was correct. Thus, we can say that the influence of Jupiter creates chaotic motion in a certain asteroid that leads to an instability in that orbit that causes it to assume one that is qualitatively different. (A snapshot of about 5000 asteroids as observed at one moment in time is shown in Figure 13.10.) It can be seen

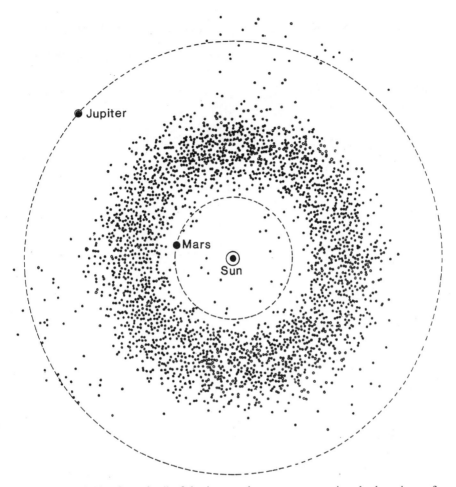

FIGURE 13.10 "Snapshot" of the inner solar system capturing the locations of the Sun, Mars, Jupiter, and approximately 5000 asteroids at one moment in time. Most asteroids orbit the Sun in a belt 1.5 a.u. wide between the orbits of Mars and Jupiter. (Courtesy of S. Ferraz-Mello, University of Sao Paulo.)

that a few move inside the orbit of Mars, and about a half dozen of these are actually inside the orbit of Earth.

There is one established case of chaotic motion in the solar system that has been demonstrated by direct observation and that is the tumbling motion of one of Saturn's satellites, Hyperion. It turns out that this satellite is not spherically symmetric. Therefore, as it tumbles, the intensity of the reflected light from the satellite changes. Figures 13.11 and 13.12 illustrate

FIGURE 13.11 Images captured by the *Voyager 2* spacecraft provide three different views of Hyperion's irregular shape. (Courtesy of NASA/Jet Propulsion Laboratory.)

the situation. Note the similarity of Hyperion's light curve to the chaotic time series shown in Figure 13.7. It is worthwhile to look for more examples of chaotic motion in the solar system to see how prevalent this is.

Is the solar system stable? What we have shown is that there is no simple answer to this question. There is no formal mathematical proof, no closed method that provides a yes or no answer. Numerical approximations

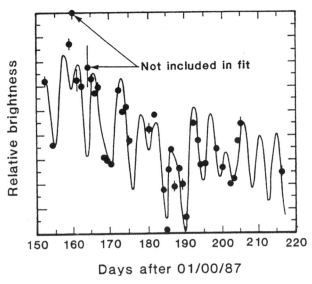

FIGURE 13.12 Plotting Hyperion's brightness each night for a period spanning more than 60 days produced an erratic pattern that no single periodic curve would fit. (Courtesy of James Klavetter.)

using a model that contains most of the major interactions between the planets over and above the influence of the Sun show that major instabilities, if they exist, have very long growth periods. At the same time, the solar system is very complex, with hundreds of thousands of minor bodies orbiting around the Sun. It has been shown that some of these orbits sometimes become chaotic and thus unstable. The solar system is unbelievably complex and rich in new problems to be investigated in the future.

For those readers who are interested in pursuing the problems discussed in this chapter in more detail, there are a number of books worth reading. A very comprehensive treatment of the three-body problem is given by V. Szebehely with special emphasis on solutions of the restricted problem of three bodies (*Theory or Orbits: The Restricted Problem of Three Bodies,* by Victor Szebehely, Academic Press, New York and London, 1967). A good discussion of stability theory for various kinds of orbits and conditions is given in several papers that are collected in a book that reports on the proceedings of the Alexander von Humboldt Colloquium on Celestial Mechanics in March 1984 (*The Stability of Planetary Systems,* edited by R. L. Duncombe, R. Dvorak, and P. J. Message, D. Reidel Publishing Company, Dordrecht and Boston, 1984). To learn more about what has come to be called chaos theory, there is a good elementary but comprehensive textbook that deals with all of the important aspects of the problem (*Nonlinear Dynamics and Chaos,* by J. M. T. Thompson and H. B. Stewart, John Wiley & Sons, New York, 1986). For those who would like to gain an understanding of the issues dealt with in this chapter without the mathematical details, there are two excellent works. One deals with chaos theory (*Chaos: The Making of a New Science,* by J. Gleick, Viking, New York, 1988) and the other with the solar system (*Newton's Clock: Chaos in the Solar System,* by I. Peterson, W. H. Freeman and Co., New York, 1993). Both of these books are easy to read, and they contain a wealth of information.

EXAMPLES

13.1. In this example, we will compute the forces acting between the bodies participating in the problem of three bodies formed by the Sun, Earth, and the Moon.

The force between the Sun and Earth is

$$F_{SE} = \frac{GM_S M_E}{r_{SE}^2} = 3.54 \times 10^{19} \text{ kg}$$

The force between Earth and the Moon is

$$F_{EM} = \frac{GM_E M_M}{r_{EM}^2} = 1.98 \times 10^{17} \text{ kg}$$

The force between the Sun and the Moon is

$$F_{SM} = \frac{GM_S M_M}{r_{SM}^2} = 4.36 \times 10^{17} \text{ kg.}$$

The following conclusions can be reached:

(a) The Moon's motion is governed by the Sun rather than by Earth since

$$F_{SM} = 2.2 F_{EM}$$

(b) Earth's motion is governed by the Sun rather than by the Moon since

$$F_{SE} = 178.8 F_{EM}$$

(c) The Sun's motion is more influenced by Earth than by the Moon since

$$F_{SE} = 81.2 F_{SM}$$

(d) The motion of the Earth–Sun system can be computed as a problem of two bodies since the effects of the Moon on the Sun and on Earth are two orders of magnitude smaller than the force between the Sun and Earth. This is the reason why the model of the restricted problem of three bodies is used only as a first approximation to compute the orbit of the Moon.

Note that when the forces were computed, the eccentricities of the orbits were neglected, and the approximation of $r_{SM} \cong r_{SE}$ was used. This relation introduces an error of 0.5% because of the variable distance between the Sun and the Moon. The effect on the force computation is 1%.

13.2. Find the location of the first colinear equilibrium point (L_1) in the Earth–Moon system using the notation of Figure 13.1.

The centrifugal force per unit mass acting on a space probe located at L_1 is

$$F_c = (L_1O)\omega^2$$

Earth's gravitational effect per unit mass on the probe is

$$F_E = \frac{GM_E}{(EL_1)^2}$$

and the Moon's gravitational effect is

$$F_M = \frac{GM_M}{(ML_1)^2}$$

The forces are balanced when

$$F_c = F_E + F_M$$

The sum of the masses of Earth and the Moon, $M_E + M_M$, will be denoted by M_T. Kepler's third law gives

$$\omega^2 = \frac{M_T G}{(EM)^3}$$

Since the center of mass is at point O,

$$OM = \frac{EM}{M_T} M_E$$

and since $L_1O = OM + ML_1$, we have

$$L_1O = \frac{EM}{M_T} M_E + ML_1$$

The distance between Earth and L_1 is

$$EL_1 = EM + ML_1$$

Now we substitute the above expressions for L_1O, ω^2, and EL_1 into the force–balance equation and write x for ML_1, l for EM, a for M_E/M_T, and b for M_M/M_T. The result is

$$x^2(la + a)(l + x)^2 - l^3ax^2 - l^3b(l + x)^2 = 0 \qquad \text{where } a + b = 1$$

The unknown in this fifth-order equation is the distance $x = ML_1$, which gives the location of the first colinear equilibrium point. All the other quantities in this equation are known. The computation of the only positive root by iteration is left to the reader.

The approximate solution is given by

$$x \approx \left[\frac{b}{3(1-b)} \right]^{1/3} l$$

The solution is $x = 61500$ km, which distance corresponds to approximately 1.54 times the circumference of Earth.

PROBLEM

13.1. Locate the coordinates of all five equilibrium points for the Earth–Moon system and for the Sun–Jupiter system (and in this way become a scientific descendant of Lagrange).

APPENDIX I

GLOSSARY

Anomalies: the angles describing the motion of a body in a reference frame as independent variables (see eccentric, mean, and true anomalies).

Anomalistic year: the mean interval between successive perihelion passages of Earth.

Aphelion: the point of a planetary orbit that is at the greatest distance from the Sun.

Apoapsis: the point of an elliptic orbit farthest from the focus.

Apogee: the point of an elliptic satellite orbit farthest from Earth.

Apsidal line: the line connecting the periapsis with the apoapsis.

Argument of latitude: the angle from the ascending node to the location of the body (measured in the orbital plane); also the sum of the argument of periapsis and the true anomaly.

Argument of periapsis: the angle between the nodes and the apsidal line in the plane of the orbit.

Ascending node: the point in the equatorial plane, or in general, in the reference plane, where the body passes from the Southern to the Northern Hemisphere.

Astrodynamics: branch of space engineering or astronautics dealing with the orbital and rigid-body motion of artificial bodies in space.

Astronautics: branch of engineering dealing with space missions.

Barycenter: the center of mass of a system of bodies.

Celestial latitude: the angle between the ecliptic and a given point, measured along the great circle.

Celestial mechanics: branch of dynamical astronomy, dealing mostly with the motion and dynamics of bodies of the solar system.

Circular restricted problem of three bodies: two bodies with large masses move on circular orbits and influence the motion of a third body with much smaller mass.

Circularization: the change of an elliptic orbit to a circular orbit by change of the velocity.

Dynamical astronomy: branch of astronomy dealing with the motion and dynamics of celestial bodies. It includes celestial mechanics, stellar dynamics, motion of binary stars, positional astronomy, etc.

Eccentric anomaly: the angle at the center of an elliptic orbit, formed by the apsidal line and the radius vector drawn from the center to the point on the circumscribing auxiliary circle from which a perpendicular to the apsidal line will intersect the orbit (see the angle E on Fig. 7.1).

Eccentricity: the distance from the center to the focus divided by the length of the semimajor axis.

Ecliptic: the mean plane of Earth's orbit around the Sun.

Ephemeris: the tabular representation of the position as a function of time of natural or artificial bodies.

Equatorial bulge: see flattening.

Equatorial satellite: a satellite orbiting in the equatorial plane of Earth.

Equinox: the intersection of the equatorial plane and the plane of the ecliptic.

Escape velocity: the velocity that results in a two-body orbit with zero velocity at infinity.

Flattening: a measure of deviation from a spherical shape, $f = (a - b)/a$, where a is the equatorial radius and b is the polar radius. Also known as oblateness or equatorial bulge, and it is applied to a body generated by the rotation of an ellipse about its minor axis.

General perturbation method: analytical solution of the differential equations describing a perturbed orbit.

Geocentric: referred to the center of Earth.

Geocentric gravitational constant: the product of the mass of Earth and the constant of gravity.

Geoid: an equipotential surface of Earth corresponding to the mean sea level of the open ocean.

Geostationary satellite: a satellite at geosynchronous altitude.

Geosynchronous altitude: the elevation above Earth's equator, where a satellite's position is fixed relative to Earth's rotation.

Gravitational constant: the factor of proportionality in Newton's law of gravity.

Gravitational harmonics: the terms in the Legendre series expansion of the gravitational potential.

Gravitational potential: a function, the derivative of which gives the gravitational force.

Heliocentric: referred to the center of the Sun.

Heliocentric gravitational constant: the product of the mass of the Sun and the constant of gravity.

Hohmann orbit: the tangential elliptic transfer orbit between two orbits of different radii or semimajor axes (see Fig. 8.4).

Hyperbolic excess velocity: the velocity above escape velocity.

Inclination: the angle between the orbital plane and the reference plane, which is the equatorial plane for planetary satellites and the ecliptic for heliocentric orbits.

Intermediate orbit: an approximation to the actual perturbed orbit (special case is the osculating orbit).

International ellipsoid: ellipsoid approximating the shape of Earth.

Invariable plane: the plane containing the center of mass of the solar system and perpendicular to the angular momentum vector of the solar system.

Isochronous stability: kinematical behavior of the disturbed path (Lyapunov's stability).

Kepler's equations: the transcendental relation between mean and eccentric anomaly.

Kepler's laws: three principles describing the motion of the planets in the solar system, generally applicable to the problems of two bodies.

Lagrangian solutions: the equilibrium solutions of the restricted problem of three bodies (see Fig. 13.1).

Lambert's theorem: a relation showing the elapsed time on a two-body orbit as a function of the chord, of the sum of the radial distances, and of the semimajor axis.

Laplace's invariable plane: see invariable plane.

Latitude: the angle between the ecliptic (or the equatorial plane) and the position vector measured at the center of the Sun (or Earth) along the great circle (or the meridian).

Libration: oscillation about equilibrium points; for instance the variation of the orientation of the Moon with respect to the Earth; also motion of bodies around the triangular equilibrium points in the restricted problem of three bodies.

Linear stability: effect of small disturbances when applied to the linearized equations of motion.

Line of nodes: the line connecting the ascending and descending nodes (see Fig. 9.1).

Longitude: the angle between the vernal equinox and the great circle, measured in the ecliptic or the angle between the Greenwich meridian and the meridian of a given position measured in the equatorial plane.

Longitude of periapsis: the sum of the angles of the longitude of the ascending node and the argument of periapsis (Fig. 9.1).

Lunar theory: a prediction of the motion of the Moon, usually analytical.

Mean anomaly: the product of the mean motion and the interval of time since pericenter passage.

Mean motion: the value of a constant angular velocity required for a body to complete one revolution.

Meridian: the great circle between the North and South poles (terrestrial and celestial) which passes through the point directly above the observer.

Newton's law of gravitation: the gravitational force between bodies is directly proportional to the product of their masses and inversely proportional to the square of their distances.

Nutation: the short-period oscillation of the pole.

Oblateness: see flattening.

Obliquity: the angle between the equatorial and orbital planes.

Obliquity of the ecliptic: the angle between the equatorial and the ecliptic planes.

Orbit: usually the path of a body with respect to another such as planetary orbit around the Sun.

Orbital stability: geometrical behavior of the disturbed orbit (Poincaré'a stability).

Orbit mechanics: branch of mechanics dealing mostly with the orbital motion of natural and artificial bodies in space.

Osculating elements: the elements of the pertubed two-body orbit which would be applicable if the pertubation would be eliminated.

Osculating orbit: the two-body orbit which would be followed if the perturbations would be turned off (see also intermediate orbit).

Parabolic velocity: see escape velocity.

Periapsis: the point of an orbit closest to the focus.

Perigee: the point of a satellite orbit closest to the Earth.

Perihelion: the point of a planetary orbit closest to the Sun.

Period: the interval of time between consecutive occasions on which the system is represented by the same vector in the phase space (same position and velocity).

Periodic motion: the motion repeats itself in equal intervals of time (recurrence of dynamical properties).

Periodic pertubations: the periodically changing effect of perturbations.

Perturbations: forces which result in deviations between the actual orbit and reference orbit, such as a two-body orbit. (See also special, general, secular and periodic pertubations.)

Phase space: combination of position and velocity coordinates (six-dimensional for the three-dimensional motion of a point mass).

Planetary theory: prediction of the motion of planets of the solar system, usually analytical.

Polar orbit: it passes over the North and South pole, its inclination is 90 degrees.

Problem of three bodies (general): the dynamics of three gravitationally interacting point masses.

Problem of three bodies (restricted): the modification of the general problem in the case when one of the three bodies, because of its small mass, is not influencing the motions of the other two bodies with much larger masses.

Problem of two bodies: the dynamics of two gravitationally interacting point masses.

Precession: the secular motion of the pole.

Rectilinear orbit: a straight-line orbit, for two bodies with unit ecentricity and zero semilatus rectum.

Regularization: the elimination of singularities from the equations of motion.

Restricted problem of two bodies: the mass of one of the bodies is much smaller than the mass of the other and the gravitational effect of the smaller body is neglected.

Sectorial harmonics: the terms in the Legendre series expansion of the gravitational potential that depend only on the longitude.

Secular perturbation: the continuously increasing effect of perturbations.

Selenocentric: referred to the center of the Moon.

Semilatus rectum: the distance between the focus and a point on the conic section measured in the direction normal to the apsidal line.

Semimajor axis: the distance from the center of an ellipse or hyperbola to an apsis.

Semiminor axis: the distance from the center of the ellipse along the line perpendicular to the apsidal line.

Sidereal time: a measure of Earth's rotation with respect to the stars.

Sidereal year: the time for Earth to complete one revolution on its orbit with respect to a fixed vernal equinox or with respect to the background stars.

Singularity: the appearance of zero distances between participating bodies, or in general, zero denominators appearing in the equations of motions.

Small divisor: the denominator appearing in perturbation analysis that approaches zero, usually because of resonance conditions.

Special perturbation method: numerical integration of the differential equations describing a perturbed orbit.

Stability: behavior of a dynamical system when disturbances are applied.

Stable motion: the effect of initially small disturbances stay below a given limit.

Stellar dynamics: branch of dynamical astronomy dealing with the motion and dynamics of stars and stellar systems such as clusters and galaxies.

Synodic period of planetary motion: the time between two successive conjunctions of two planets, as observed from the Sun.

Synodic period of satellite motion: the time between two successive conjunctions of the satellite with the Sun, as observed from the satellite's planet.

Synodic system: in general, a coordinate system rotating around the center of mass of the participating bodies.

Terrestrial harmonics: terms in the Legendre series expansion of Earth's gravitational potential.

Terrestrial latitude: the angle between the equatorial plane and a given point, measured along the meridian.

Tesseral harmonics: the terms in the Legendre series expansion of the gravitational potential that depend on the latitude and longitude) J_k^j, $j \neq k$).

Time of perigee passage: the time when a satellite passes closest to Earth.

Topocentric: referred to a point on the surface of Earth.

Trajectory: usually the part of an orbit such as a missile's or rocket's path.

True anomaly: the angle at the focus (nearest the pericenter) between the apsidal line and the radius vector (drawn from the focus to the orbiting body). (See angle f in Figs. 4.4 and 7.1.)

Unstable motion: the effect of initially small disturbances increase above a given limit.

Vernal equinox: the direction where the Sun crosses the equatorial plane from south to north in its apparent motion along the ecliptic (its apparent longitude is zero); the ascending node of the ecliptic on the equatorial plane.

Zonal harmonics: the terms in the Legendre series expansion of the gravitational potential that depend only on the latitude.

APPENDIX II

PHYSICAL CONSTANTS

The reader is reminded that the following constants change as better measurements and observations become available, but their value will never be known "exactly." Most of the numbers presented are based on the 1997 edition of the Astronomical Almanac issued by the Nautical Almanac Office, U.S. Naval Observatory and published by the U.S. Government Printing Office, Washington, D.C. These numbers do not always agree completely with those given by other sources, such as the International Astronomical Union and the International Association of Geodesy. For instance the IAG value for Earth's equatorial radius is 6378136 ± 1 m, which constant in our table is 6378.14 km.

Some of the entries are redundant since the reader certainly can compute the constant gravity (G) if the geocentric gravitational constant (GM_E) and the value of Earth's mass (M_E) are given. The reason for furnishing such redundant values is to facilitate the solution of actual problems in space dynamics. Some of the constants for Earth are given in the table (rounded-off values) and also are listed with the presently existing highest accuracy, showing the error limits.

The effects of uncertainties in orbit mechanics are discussed by G. B. Westrom's article, which appeared in the proceedings of a symposium on "Space Trajectories," published by Academic Press, New York (1960), and by a fascinating paper by Sir James Lighthill, "The Recently Recognized Failure of Predictability in Newtonian Dyanmics," *Proceedings of the Royal Society of London,* Vol. A407, pp. 35–50, 1986.

For additional information see the above-mentioned Astronomical Almanac or its Explanatory Supplement (1961, revision published in 1988). A. E. Roy's *Orbital Motion,* listed in the references, also has many of the useful constants.

Table of Constants of the Planets

Planet	R_e (km)	M $(10^{24}\,\text{kg})$	J_2 (10^{-3})	i (degrees)	a (a.u.)	e
Mercury	2,439	0.33022	—	7.006	0.3871	0.2056
Venus	6,052	4.8690	0.027	3.395	0.7233	0.0067
Earth	6,378.14	5.9742	1.02863	0.001	1.0000	0.0167
Mars	3,397.2	0.64191	1.964	1.851	1.5237	0.0933
Jupiter	71,398	1,899.2	14.75	1.305	5.2030	0.0482
Saturn	60,000	568.56	16.45	2.486	9.5281	0.0542
Uranus	25,400	86.978	12	0.663	19.1829	0.0459
Neptune	24,300	102.98	4	1.769	30.0796	0.0101
Pluto	2,500	0.663	—	17.142	39.3396	0.2462

Notes

R_e: equatorial radius

M: mass

J_2: second zonal gravitational harmonic

i: inclination to the mean ecliptic

a: semi-major axis

e: eccentricity

The heliocentric osculating orbital elements given in this table are referred to the mean ecliptic and equinox of J2000 and are given for February 9, 1988.

Additional Constants of the Solar System

Equatorial radius of the Sun: 696000 km

Mass of the Sun: 1.9891×10^{30} kg

Heliocentric gravitational constant: $GM_s = 1.3271244 \times 10^{20}$ m³/s²

Astronomical Unit: 1.49597870×10^8 km

$$\frac{\text{Mass of Sun}}{\text{Mass of Jupiter}} = 1047.355$$

$$\frac{\text{Mass of Sun}}{\text{Mass of Earth}} = 332{,}946$$

$$\frac{\text{Mass of Sun}}{\text{Mass of Earth} + \text{Moon}} = 328{,}900.5$$

Equatorial radius of Earth: $R_e = (6{,}378{,}136 \pm 1)$ m
Polar radius of Earth: $R_p = (6{,}356{,}751 \pm 1)$ m
Flattening of Earth: $f = 3.35281 \times 10^{-3}$
Second zonal harmonic of Earth: $J_2 = (108263) \times 10^{-8}$
Period of Earth's revolution (one sidereal year): 356.25636 days
Anomalistic year: 365.25964 days
Angular velocity of Earth's rotation: $\omega = (7.292115 \pm 10^{-7}) \times 10^{-5}$ rad/s
Period of Earth's rotation (one sidereal day): $T = 23$ h 56 min 4.1 s
Obliquity of the ecliptic: $23°26'21''.448$ at standard epoch 2000.
Geocentric gravitational constant: $GM_E = (3986004.48 \pm 0.03) \times 10^8$
    m^3/s^2

$$\frac{\text{Mass of Moon}}{\text{Mass of Earth}} = 0.0123000$$

Mean radius of the Moon: 1738 km
Mass of the Moon: 7.3483×10^{22} kg
Semimajor axis of the lunar orbit: 384400 km
Moon's orbital period (sidereal): 27.321661 days
Eccentricity of the lunar orbit: 0.0549
Constant of gravitation: $G = 6.672 \times 10^{-11}$ $\text{m}^3/\text{kg s}^2$
Speed of light: 299,792,458 m/s

ANNOTATED LIST OF MAJOR REFERENCE BOOKS

The preparation of such lists presents many difficult choices. Inclusion of too many items will not help readers, but anything important left out will be to their definite disadvantage. The notes added might help the readers to decide how to spend their time and where to turn to satisfy their curiosity. A few books emphasizing the historical aspect are also included to allow the readers to balance science and humanities.

H. Alfven and G. Arrhenius (Eds.), *Evolution of the Solar System.* NASA SP-345, U.S. Government Printing Office, Washington, D.C., 1976. This work is an excellent and also comprehensive description of the state of knowledge of the Solar System following the first decade of exploration with spacecraft carrying out interplanetary missions.

E. N. da C. Andrade, *Sir Isaac Newton.* Macmillan, New York, 1954. Combination of Newton's life, works, personality, human dimension and detailed biography.

V. I. Arnold, *Mathematical Methods of Classical Mechanics.* Nauka, Moscow, 1974. Translation published by Springer-Verlag, New York, 1978. Modern mathematical approach to dynamics, including Newtonian, Lagrangian and Hamiltonian formulations. Classical mechanics is related to areas of mathematics such as Riemannian geometry, Kolmogorov's theorem, Lie groups, mapping theorems, etc. This is an advanced text book used by Arnold teaching classical mechanics at Moscow State University.

R. M. L. Baker and M. W. Makemson, *An Introduction to Astrodynamics.* Academic Press, New York, 1960. Second revised edition 1967. Easy to read, written in clear style, mostly for engineers. Many useful exercises concerning space dynamics.

V. R. Baker, *The Channels of Mars*. The University of Texas Press, Austin, 1982. This book is focused on the topography of Mars with special emphasis on features that might have a bearing on the presence of liquid water on the planet in the past. It has many spectacular illustrations.

R. R. Bate, D. D. Mueller and J. E. White, *Fundamentals of Astrodynamics*. Dover, New York, 1971. This textbook is for introductory engineering courses, emphasizing the systems engineering approach, the practical aspects of orbit determinations, lunar and interplanetary trajectories (with patched conic approximations) and ballistic missile trajectories. The "historical digressions" are short and interesting as are the many exercises.

R. H. Battin, *An Introduction to the Mathematics and Methods of Astrodynamics*. American Institute of Aeronautics and Astronautics, New York, 1987. A major reference work from the basic two-body problem to highly sophisticated problems of engineering space dynamics with many fascinating personal remarks describing the U.S. space program of which the author was one of the major contributors. The large amount of material included and many mathematical details require devoted readership. Battin's earlier book, entitled *Astronautical Guidance,* McGraw Hill, Inc. (1964) is another volume of considerable importance to space engineering. The Apollo guidance system is discussed in detail by the author who was the director of the space guidance analysis division of this project.

A. Beer and K. A. Strand (editors), *Copernicus. Vistas in Astronomy.* Vol. 17, Pergamon Press, New York, 1975. This volume represents the proceedings of a conference held in 1972 in Washington, D.C. to commemorate the 500th anniversary of the birth of Copernicus. It highlights Copernicus' ideas and place in the history of celestial mechanics.

G. D. Birkhoff, *Dynamical Systems*. American Mathematical Society, New York, 1927. Advanced theoretical treatment emphasizing the qualitative aspect of dynamics. Main subjects are stability, periodic orbits, problem of three bodies and integrability.

D. Brouwer and G. M. Clemence, *Methods of Celestial Mechanics*. Academic Press, New York, 1961. This advanced reference and textbook is one of the classics in the field of celestial mechanics. Astronomically oriented and dedicated readers will find a large amount of very useful information.

E. W. Brown, *An Introductory Treatise on the Lunar Theory.* University Press, Cambridge, 1896. Also Dover, New York, 1960. A clear description of the problem of the motion of the Moon with several approaches to the solution. The major methods (Laplace, de Pontecoulant, Hansen, Delaunay, Hill, etc.) are explained and compared. This is a major reference book rather than an introductory treatment.

E. Burgess, *Uranus and Neptune: The Distant Giants*. Columbia University

Press, New York, 1988, and *Far Encounter: The Neptune System.* Columbia University Press, New York, 1991. These books are both well illustrated and highly readable accounts of the exploration of Uranus and Neptune.

P. Cattermole, *Venus: The Geological Story.* The Johns Hopkins University Press, Baltimore, 1994. This is a richly illustrated volume using the radar images obtained by Magellan. The striking pictures in this book provide evidence of tectonic motion, volcanic activity and many other phenomena. This is recommended reading for anyone who wants to learn more details about the planet.

C. V. L. Charlier, *Die Mechanik des Himmels.* Vols. 1 and 2, Viet and Co., Leipzig, 1902–1907. Mathematically oriented, highly advanced, but quite readable German text.

J. Chazy, *Mécanique Céleste.* Presses Universitaires de France, Paris, 1953. Delightfully short and concise French text on a rather advanced level, discussing canonical transformations, variational equations, perturbation theories, etc.

G. A. Chebotarev, *Analytical and Numerical Methods of Celestial Mechanics.* Nauka, Moscow, 1965. English translation by L. Oster, American Elsevier, New York, 1967. Advanced astronomically oriented, clearly written, graduate textbook. Planetary theories, lunar theories, study of minor planets, satellites and comets.

P. E. Cleator, *Rockets Through Space: The Dawn of Interplanetary Travel.* Simon and Schuster, New York, 1936. This book describes the development of rockets after World War I. It is important because it is oriented toward the development of rockets for space flight and it contains many original mission profiles and calculations that outline the energy levels and velocity increments that must be achieved to carry out interplanetary and Earth orbital flight. The book is very well written.

N. Copernicus, *De Revolutionibus Orbium Coelestium. (The Revolution of Heavenly Spheres.)* Norimbergae (Nurnberg), apud Ioh. Petreium, 1543. This book, written in Latin, was published under the Imprimatur of the Bishop of Frauenberg. It is the first detailed description of the Solar System with the Sun placed at the center. The book was published after the death of Copernicus in 1543.

J. M. A. Danby, *Fundamentals of Celestial Mechanics.* Macmillan, New York, 1962. The use of vectors can simplify some of the equations and ideas of celestial mechanics as demonstrated in this basically introductory book with several well-selected exercises.

R. Deutsch, *Orbital Dynamics of Space Vehicles.* Prentice-Hall, Englewood Cliffs, New Jersey, 1963. Advanced text emphasizing the theoretical foundations and outlining the solution techniques used in celestial mechanics. The subjects treated in some detail, besides the basics, are orbit determination, analytical dynamics as applied to general perturbation techniques, and modern topological research related to the problem of three bodies.

G. N. Duboshin, *Celestial Mechanics, Fundamental Problems and Methods.* Nauka, Moscow, 1968 and *Celestial Mechanics, Analytic and Quantitative Methods.* Nauka, Moscow, 1978. These basic and important books in our field are available only in their original Russian editions.

D. Dubyago, *The Determination of Orbits.* Nauka, Moscow, 1949. English translation by R. D. Burke, G. Gordon, L. N. Rowell and F. T. Smith, Macmillan, New York, 1961. Book is basically for astronomers, discussing the determination of orbits of minor planets, comets and meteors. Many excellent examples.

K. A. Ehricke, *Space Flight.* Vol. I, Environment and Celestial Mechanics, 1960. Vol. II, Dynamics, 1962. Vol. III, Operations, 1964. D. Van Nostrand, Princeton, New Jersey. Space missions and system analysis are emphasized in these easy to read volumes, directed to astronautical engineers, with details on ballistics and powered flights. Many well selected examples and references.

B. Erdi, *Egi Mechanika (Celestial Mechanics).* Vols. 1–3, Eötvös University, Budapest, 1972–74. Subjects emphasized are orbit determination, general perturbations, lunar theory, dynamics of artificial satellites. This clearly formulated basic text is available only in Hungarian.

P. R. Escobal, *Methods of Orbit Determination.* R. E. Krieger, Huntington, New York, 1965. Oriented to aerospace engineers and applied astrodynamicists, this highly readable book offers a large amount of very useful information with many examples and detailed computational algorithms.

E. C. Ezell and L. N. Ezell, "On Mars: Exploration of the Red Planet 1958–1978." NASA SP-4212, U.S. Government Printing Office, Washington, D.C., 1982. This books contains a very good nontechnical account of the exploration of Mars up to the two *Viking* missions in 1976. It is very well written and it is recommended for students who wish to gain background knowledge about Mars without getting into the technical details. There are many interesting illustrations.

E. Finlay-Freundlich, *Celestial Mechanics.* Pergamon Press, New York, 1958. Delightfully short and clear. The problem of two bodies treated in the introductory chapter which is followed by advanced dynamical astronomy.

P. M. Fitzpatrick, *Principles of Celestial Mechanics.* Academic Press, New York, 1970. Emphasis on artificial satellites, including rigid body rotational motion. Advanced and often mathematically oriented treatment with many exercises.

K. F. Gauss, *Theoria Motus Corporum Coelestium in Sectionibus Conicis Solem Ambientium (Theory of the Motion of the Heavenly Bodies Revolving around the Sun in Conic Sections).* Hamburg, 1809. English translation by C. H. Davis, Little, Brown, Boston, 1957. One of the classics of theoretical and computational celestial mechanics, emphasizing orbit determination (such as orbits from three observations), method of least squares, orbits of Ceres, Pallas and Juno, etc.

T. Gehrels, (Ed.), *Jupiter.* The University of Arizona Press, Tucson, 1976. This is an excellent collection of papers reporting on the results of *Pioneer 10* and *11* fly-bys of the planet. This book is intended for a technical or scientific expert on the subject.

J. Gleick, *Chaos: The Making of a New Science.* Viking, New York. 1988. This is a highly readable and imaginative account of how chaos theory has been developed. It is strongly recommended for someone who wants to learn about the topic without becoming enmeshed in mathematical details.

R. H. Goddard, "A Method for Reaching Extreme Altitudes." Smithsonian Institution, Miscellaneous Collection LXXI, No. 2, p. 69 (1919) and "Report to the Smithsonian Institution Concerning Further Developments of the Rocket Method of Investigating Space," reprinted in: E. C. Goddard and G. E. Pendray, (Eds.), *The Papers of Robert H. Goddard.* McGraw-Hill, New York, 1970. These papers represent Goddard's seminal work on rocket propulsion. The rocket equations are derived and experimental results are reported in quantitative detail both for solid and liquid fueled rockets. In the second paper Goddard describes the use of rockets to reach space above the atmosphere.

H. Goldstein, *Classical Mechanics.* Addison-Wesley, Reading, Mass., 1950. Introductory (compared to Whittaker's book), easy to read, clear, basic reference text.

Y. Hagihara, *Celestial Mechanics.* Vols. 1 and 2, Massachusetts Institute of Technology Press, Cambridge, Mass., 1970–1972; Vols. 3–5, Japan Society for Promotion of Science, Tokyo, Japan, 1974–1976. These volumes represent the encyclopedia of celestial mechanics. The reader will find everything discussed in considerable detail, even those subjects or approaches which are only indirectly connected with celestial mechanics. This major reference book is comprehensive, clear and connects astronomy with mathematics.

S. W. Hawking and W. Israel, *300 Years of Gravity.* Cambridge University Press, London, 1987. Several contributors discussing mostly physics since Newton: cosmology, relativity, black holes, superstring unification and quantum theory.

P. Herget, *The Computation of Orbits.* Edwards Brothers, Ann Arbor, Michigan, 1948. This astronomically oriented, easy to read, advanced text specializes in orbit determination. The author's and his associates' humor will be enjoyed by those scholars who carefully translate the Greek language subtitles of the chapters. For instance: Chapter 2. Problems in Spherical Astronomy. (Leave hope behind all ye who enter here.) Chapter 6. Improvement of the Orbit. (If at first you do not succeed try, try again.) Chapter 7. Special Perturbations. (These numbers laid end to end would reach to insanity.)

S. Herrick, *Astrodynamics.* Vols. 1 and 2, Van Nostrand Reinhold, London, 1971–1972. Basic text with applications to space engineering and astronomy. Some of the unconventional notations require careful attention. Principle subjects are special and general perturbation theories, orbit determination and universal variables. Many examples with details.

G. Hunt and P. Moore, *Saturn*. Rand McNally, New York, 1982. This is a very good study of Saturn and its rings and satellites. This book is recommended for anyone who is interested in expanding his or her knowledge of Saturn. It is a well written book.

R. Jastrow and S. I. Rasool (Eds.), *The Venus Atmosphere*. Gordon and Breach, New York, 1969. This is a detailed and comprehensive description of the atmosphere of Venus after the first fly-bys by *Mariner* and *Venera* spacecraft and the first atmospheric entry by *Venera 4*. The book is intended for those students who are interested in studying planetary science.

M. H. Kaplan, *Modern Spacecraft Dynamics and Control*. J. Wiley, New York, 1976. This is an excellent undergraduate textbook on the subject. It is easy to read and has many good problems.

D. G. King-Hele, *Satellites and Scientific Research*. Routledge and Kegan Paul, London, 1962 and *Theory of Satellite Orbits in an Atmosphere*. Butterworths, London, 1964. One of the important contributors to the still unsolved problem of atmospheric effects on satellite motion. See also his Technical Memorandum No. 212 of the Royal Aircraft Establishment, entitled "A View of Earth and Air," 1974.

A. Koestler, *The Sleepwalkers*. Macmillan, New York, 1959. History of science from about 3000 B.C. to Newton. Book emphasizes history rather than science, and the "cold war" between humanities and science.

J. Kovalevsky, *Introduction a la Méchanique Céleste*. Librairie Armand Colin, Paris, 1963. English translation, D. Reidel, Dordrecht, Holland, 1967. This short and well-written book emphasizes perturbation theories and the problem of artificial satellites. In the author's opinion, "the construction of increasingly more accurate analytical theories remains the central task of celestial mechanics."

J. Lagrange, *Oeuvres*. 14 Volumes, Gauthier-Villars, Paris, 1867–1892 and *Méchanique Analytique*. Veuve Desaint, Paris, 1788. Author's infatuation with equations makes his work somewhat difficult to read. The brilliance of the presentations and the analytical manipulations show clearly Lagrange's mastery of his subjects and will fill the reader with admiration of these volumes written in French.

P. S. Laplace, *Traité de Mécanique Céleste*. Vols. 1–3, Duprat, Paris, 1799–1802; Vol. 4, Courrier, Paris, 1805; Vol. 5, Bachelier, Paris, 1823–1825. English translation by N. Bowditch, Volumes 1–4, Chelsea, New York, 1829–1839. This major classic of celestial mechanics is directed to advanced, astronomically oriented readership. The comments and detailed explanations in the English translation are of considerable help.

A. B. Lerner, *Einstein and Newton*. Lerner, Minneapolis, Minnesota, 1873.

Lives, backgrounds, and accomplishments compared via many valuable quotations from correspondences.

S. W. McCuskey, *Introduction to Celestial Mechanics*. Addison-Wesley, Reading, Massachusetts, 1963. This easy to read, clear, introductory text emphasizes the basic principles and offers well selected examples.

E. D. Miner, *Uranus: The Planet, Rings and Satellites*. Ellis Harwood, New York, 1990. This book is a detailed and comprehensive study of Uranus and her system. It is intended for graduate students but the material is presented in such a way that it is also useful for undergraduates as well.

J. Moser, *Stable and Random Motions in Dynamical Systems with Special Emphasis on Celestial Mechanics*. Princeton University Press, Princeton, NJ, 1973. The author of this mathematical exposition is the third member of the famous KAM trio (Kolmogorov, Arnold, Moser). The book concentrates on the convergence problem of the series solutions, the role of small divisors, non-integrability, ergodic motions, quasi-periodic behavior, etc. With patience and some analytical background the reader will enjoy this version of Moser's Hermann Weil lecture series delivered at the Institute for Advanced Study in 1972.

F. R. Moulton, *An Introduction to Celestial Mechanics*. Macmillan, New York, 1960. This book is an easy to read classic with astronomical orientation. The historical remarks and well-selected references are most informative.

I. Newton, *Philosophiae Naturalis Principia Mathematica (The Mathematical Principles of Natural Philosophy)*. Royal Society of London, 1687. Translation by A. Motte (1729) edited by F. Cajori, University of California Press, Berkeley, California, 1946. Difficult to read because of the author's well known insistence of using classical geometry to treat dynamical problems. It is reasonable to assume that Newton derived many of his theorems by calculus but presented the proofs with geometrical tools. To quote Laplace: "The *Principia* is pre-eminent above any other production of human genius." The First Book is on theoretical mechanics, containing his laws of motion, law of gravitation, and problem of two bodies. The Second Book is on hydrodynamics. The Third Book forms the major part of the *Principia* and treats the dynamics of satellites and planets, including the problem of mass determination, flattening, and precession of the equinoxes. Lunar theory, theory of tides and orbits of comets conclude the Third Book.

H. Oberth, *Die Rakete zu dem Planetenraum*. Oldenburg, Munich, 1923, reprinted by Uni-Verlag, Nurnberg, 1960 an d*Wege zur Raumschiffahrt*. 1929, reprinted by Kriterion, Bucharest, 1974. These are popular works that were the first to attract a large circulation. Oberth's books inspired the first really popular space related science fiction movie, Fitz Lang's "Die Frau Im Mond" ("The Girl on the Moon") which was premiered in 1929.

F. I. Ordway and M. R. Sharpe, *The Rocket Team*. MIT Press, Cambridge, 1982. This is an excellent history of rocketry from the end of World War II to the

Apollo Project. It is recommended as good background reading so that students can understand why large rockets were built.

L. A. Pars, *Treatise on Analytical Dynamics*. Heinemann, London, 1965. Excellent presentation of dynamics, emphasizing the principle of least action. Many references and examples.

I. Peterson, *Newton's Clock: Chaos in the Solar System*. W. H. Freeman, New York, 1993. This book is highly recommended for all students of celestial mechanics. It is a sophisticated yet readable account of the development of our knowledge about the workings of the solar system. A limited understanding of mathematics is required to get the benefit of this important work.

H. C. Plummer, *An Introductory Treatise on Dynamical Astronomy*. Cambridge University Press, London, 1918. Reprinted by Dover, New York, 1960. This book is strongly oriented to dynamical astronomy in a clear and understandable style, on an advanced level. In addition to the basic problems of two and three bodies, considerable details are offered on planetary theory, lunar theory, determination of orbits, orbits of binary stars, perturbation methods, etc.

H. Poincaré, *Les Méthodes Nouvelles da la Mécanique Céleste*. Gauthier-Villars, Paris, 1892–1899. Reprinted by Dover, New York, 1957. English translation NASA-TTF-450, 1967. It is difficult to read because of the author's concise style, but it has many brilliant ideas and suggestions for future research in celestial mechanics, especially along mathematical lines.

H. Pollard, *Mathematical Introduction to Celestial Mechanics*. Prentice-Hall, Englewood Cliffs, New Jersey, 1966. The author shows how a mathematician might write an easy to read and clear book. Some of the subjects emphasized in the book might be of more interest to mathematicians than to astronomers or to engineers.

I. Prigogine, *From Being to Becoming*. W. H. Freeman, San Francisco, 1980. Book connects the "classical period" of dynamics with the modern aspects of statistical mechanics. The reader will find a refreshing view of instability, non-integrability, and the complexities which uncertainties can create in the physical sciences.

J. H. Rogers, *The Giant Planet Jupiter*. Cambridge University Press, Cambridge, MA, 1995. This is an excellent and highly readable account of our modern knowledge of Jupiter. This book is strongly recommended for anyone who wants to learn more about Jupiter and its satellite system.

A. E. Roy, *Orbital Motion*. Adam Hilger, Bristol, 1978. This book is an easy to read, excellent text, offering combinations of astronomical and space research applications. Many advanced subjects are covered and several useful exercises are included.

Y. Ryabov, *Celestial Mechanics*. Foreign Languages Publishing House, Moscow, 1959. Also Dover, New York, 1961. This semi-popular book is simple and presents basic ideas without much mathematics and with many numerical explanations. For the layperson the book is heavy reading, but the undergraduate astronomy or engineering student will consume it with pleasure in a short time with considerable benefit.

C. L. Siegel and J. K. Moser, *Lecture on Celestial Mechanics.* Springer-Verlag, Berlin, 1971. This is a readable, mathematically oriented book, based on Siegel's book published in 1956. It concentrates on the problem of solution of differential equations of celestial mechanics and on the problem of three bodies. It is based on the works of Poincaré, Sundman, Birkhoff and Wintner.

W. M. Smart, *Celestial Mechanics.* Longmans, Green, London, 1953. One of the few easy to read, advanced books on celestial mechanics, with astronomical orientation. Almost one third of the book deals with lunar theories. Few references and no exercises are given but complete and clear explanations of the basic theoretical aspects are offered.

T. E. Sterne, *An Introduction to Celestial Mechanics.* Interscience Publishers, New York, 1960. This short and clear test is for advanced level readers. It contains the author's original contributions to the analysis of the dynamics of artificial satellites.

E. L. Stiefel and G. Scheifele, *Linear and Regular Celestial Mechanics.* Springer-Verlag, Berlin, 1971. Book describes regularizing transformations which result in linear differential equations for the problem of two bodies. Generalizations are discussed from two to three dimensions, using canonical theories. Several results are shown, concerning the numerical advantages of regularization.

K. Stumpff, *Himmelsmechanik.* Vols. 1–3, Deutscher Verlag der Wissenschaften, Berlin, 1959–1974. Clear, easy to read treatment emphasizing orbit determination, the problem of three bodies and perturbation theories. Many details make reading this book, written in German, a rather lengthy but definitely worthwhile project.

G. P. Sutton, *Rocket Propulsion Elements: An Introduction to the Engineering of Rockets.* Sixth Edition, J. Wiley, New York, 1992. This is an excellent treatment of the science and engineering of rocket propulsion. It is written at the graduate level but the introductory chapters are useful for undergraduate students.

V. Szebehely, *Theory of Orbits.* Academic Press, New York, 1967. This advanced reference text discusses the restricted problem of three bodies with theories and applications concerning periodic orbits, space trajectories, stability and dynamical astronomy.

L. G. Taff, *Celestial Mechanics: A Computational Guide for the Practitioner.* J. Wiley, New York, 1985. This highly readable, often informal text is prepared for advanced courses, emphasizing orbit determination and computational approaches.

F. Tisserand, *Traité de Mécanique Céleste.* Vols. 1–4, Gauthier-Villars, Paris,

1889–1896. This book, written in French, is a remarkable advanced text directed to astronomically oriented readers. It is more readable than some of the other classics, such as Poincaré or Laplace. Contains discussions of general perturbation methods, figures of celestial bodies and their rotational motions, theory of the motion of the Moon, theory of motion of Jupiter's and Saturn's satellites and of minor planets and comets.

J. M. T. Thompson and H. B. Stewart, *Non-Linear Dynamics and Chaos.* J. Wiley, New York, 1986. This book is an excellent quantitative treatment of the subject of chaos theory and related topics. It is intended for graduate students but some of the material is appropriate for undergraduates as well.

W. T. Thomson, *Introduction to Space Dynamics.* J. Wiley, New York, 1961. More along the lines of gyrodynamics, optimization, flexible missiles, etc. than celestial mechanics. Many examples of practical importance.

H. C. Urey, *The Planets: The Origin and Development.* Yale University Press, New Haven, 1952. This books provides a detailed description of the Solar System before the era of exploration with spacecraft. It provides an excellent starting point for students who want to learn more about how our knowledge about the planets accumulated. The book is not easy to read.

E. T. Whittaker. *A Treatise on the Analytical Dynamics of Particles and Rigid Bodies.* Cambridge University Press, London, 1904. This is one of the excellent books from which one can learn advanced dynamics. Many applications to celestial mechanics are treated in a clear style.

W. E. Wiesel, *Spaceflight Dynamics.* McGraw-Hill, New York, 1995. This book is a modern treatment of both spacecraft dynamics and orbital mechanics. There are a number of elegant derivations in this book that are worth studying in detail.

J. N. Wilford, *Mars Beckons.* Alfred A. Knopf, New York, 1990. John Noble Wilford is one of the most distinguished science writers working today, having received two Pulitzer Prizes for his work. Perhaps the best way to describe this book is to cite the statement on the dust jacket which talks of "the mysteries, the challenges and the expectations of our next great adventure in space." This book is very well written and it is highly recommended.

A. Wintner, *The Analytical Foundations of Celestial Mechanics.* Princeton University Press, New Jersey, 1941. The author wishes to emphasize "analytical foundations" in the title pointing out that "it is almost forgotten how much the theory of analytic functions owes to the elementary problem of two bodies." This book was written by and is appreciated by "ϵ-trained mathematicians" more than by engineers and astronomers. Excellent set of references are given, and it is carefully pointed out that the traditional references to the origin of the fundamental mathematical notions in analytical dynamics are often incorrect.

INDEX